T0331600

The Rise of Modern Science Explained

For centuries, laymen and priests, lone thinkers and philosophical schools in Greece, China, the Islamic world and Europe reflected with wisdom and perseverance on how the natural world fits together. As a rule, their methods and conclusions, while often ingenious, were misdirected when viewed from the perspective of modern science. In the 1600s, thinkers such as Galileo, Kepler, Descartes, Bacon and many others gave revolutionary new twists to traditional ideas and practices, culminating in the work of Isaac Newton half a century later. It was as if the world was being created anew. But why did this re-creation begin in Europe rather than elsewhere? This book caps H. Floris Cohen's career-long effort to find answers to this classic question. Here he sets forth a rich but highly accessible account of what, against many odds, made it happen and why.

H. FLORIS COHEN is Professor of Comparative History of Science at Utrecht University, where he serves as the Editor of the History of Science Society. He first explored the rise of modern science in *Quantifying Music*, examined how other historians conceived of the rise of modern science in *The Scientific Revolution: A Historiographical Inquiry* and solved the problem of how modern science arose in *How Modern Science Came into the World: Four Civilizations, One Seventeenth-Century Breakthrough*.

The Rise of Modern Science Explained: A Comparative History

H. FLORIS COHEN
Utrecht University

CAMBRIDGE
UNIVERSITY PRESS

CAMBRIDGE
UNIVERSITY PRESS

University Printing House, Cambridge CB2 8BS, United Kingdom

One Liberty Plaza, 20th Floor, New York, NY 10006, USA

477 Williamstown Road, Port Melbourne, VIC 3207, Australia

4843/24, 2nd Floor, Ansari Road, Daryaganj, Delhi - 110002, India

79 Anson Road, #06-04/06, Singapore 079906

Cambridge University Press is part of the University of Cambridge.

It furthers the University's mission by disseminating knowledge in the pursuit of education, learning and research at the highest international levels of excellence.

www.cambridge.org
Information on this title: www.cambridge.org/9781107120068

Originally published in Dutch by Uitgeverij Prometheus / Bert Bakker as *De herschepping van de wereld: het ontstaan van de moderne natuurwetenschap verklaard* (9789035133471) in 2007

The translation (subsidised by the Dutch Foundation for Literature) is a co-production by Chris Emery and the author

First published 2015

A catalogue record for this publication is available from the British Library

Library of Congress Cataloging in Publication data
Cohen, H. Floris.
[Herschepping van de wereld. English]
The rise of modern science explained : a comparative history / H. Floris Cohen, Utrecht University.
pages cm
Originally published in Dutch as: De herschepping van de wereld : het ontstaan van de moderne
natuurwetenschap verklaard (Amsterdam : Uitgeverij Bert Bakker, 2007).
"The translation (subsidised by the Dutch Foundation for Literature) is a coproduction by Chris Emery and the author."
Summary: Covers scientific discovery from approximately 1500-1699.
Includes index.
ISBN 978-1-107-12006-8 (Hardback) – ISBN 978-1-107-54560-1 (pbk.) 1. Science–History.
2. Science–History–16th century. 3. Science–History–17th century. I. Title.
Q125.2.C6413 2015
509–dc23
2015011108

ISBN 978-1-107-12006-8 Hardback

N ederlands
letterenfonds
dutch foundation
for literature

The publisher gratefully acknowledges the support of the Dutch Foundation for Literature.

Contents

Introduction: the Old World and the New

If we had been born a couple or more centuries ago, there is a good chance that we would have been poor, very poor indeed. Our lives would have been spent working on the land, with little or no prospect of change. We would produce large families, but few of our children would outlive us. In any case, we ourselves wouldn't expect to live much beyond about 45. We would call a hovel our home, and heat it in winter with whatever sticks of wood that we were able to collect. Other comforts would have to be paid for either in kind or with the few pennies we had managed to save. Apart from everyday conversation, crying children and the noise of poultry and livestock, we would have lived amidst silence, interrupted on occasion by a thunderclap, communal song, the drums and trumpets of passing soldiers, maybe a lonely church bell tolling every so often. Most of us would have believed without question in the literal existence of spirits or gods or one God as the guiding or even the all-determining power in life and even more so after death.

In short, we would have been living in what is sometimes called the Old World, as distinct from the modern New World which you and I inhabit and which has made us rich and next year may make us richer still. Ours is a world where goods are readily available everywhere; however many we may own today we can always obtain newer and more up-to-date versions of them tomorrow. We are living longer and we are dying of mostly different diseases. Noise surrounds us wherever we go. We are parents of a few carefully planned and properly vaccinated children, who are likely to live even longer than we will. Many of those among us who still regularly go to church are no longer inclined to take the texts recited there in their literal sense. Our everyday behaviour is oriented towards this life – however hard we may try we find it difficult to imagine what life after death might possibly be like. When we travel we usually reach our destination by air, rail or motor car within hours instead of after days or weeks on horseback or on board a pitching barque, more concerned about tailbacks or punctures than ambush by pirates or robbers.

The modern way of life outlined above is everyday reality for most of us in the West. For the majority of the world's population it is a different matter or, rather, it is still a different matter. The minority for whom the picture of the modern world just sketched has already become a reality is growing by the day. And the diminishing majority who hardly take part in it yet do not just aspire to the material advantages that we may call ours: their aspiration has become perfectly realistic. Even if the world's poor do not expect it for themselves, they do so for their children or their children's children. Within a few generations the leap from the Old to the New World seems to be attainable for all.

This of course raises the question of what made the leap possible. When, where, how and due to what was it first made?

When and where it all started are easy to answer: indeed, the modern world has its roots in the West, and the first signs manifested themselves in Britain around 1780. As a result, within a century, the face of Europe and the United States changed beyond recognition.

Precisely how the process of modernisation developed and in particular how a New World was actually able to detach itself from the Old, and why this turning point in history took place specifically in European civilisation rather than in China or India or the Islamic world – these are for a historian the truly big questions. The past hundred years or so have seen a succession of studies devoted to untying this tangled knot which, in its full complexity, I shall leave tied here. In this book I confine myself to resolving one specific aspect of the question – an aspect that is often overlooked or cut short but is nevertheless crucial. Directly or indirectly, at every point in the range of contrasts that I listed above, we encountered modern science.

Take the contrast between modern unbelief and the pre-modern literal belief in a quite vividly imagined hereafter. What is at issue here is our irreversible awareness of abstract natural laws, operating according to fixed rules, in precisely defined circumstances – laws of the kind that have continued to characterise nature-knowledge ever since Newton. These laws have made the notion of a deity concerned with our personal welfare highly problematic. Whether modern science actually *imposes* some 'scientific world-view' is a doubtful matter. It is nonetheless fairly obvious that modern science is at odds in important ways with the broad conception of the world that comes with the traditional world religions.

Another contrast listed above, between pre-modern silence and modern noise, does not so much reflect the tension between a pre-scientific world-view and one formed at least in part by modern science. Here the contrast is between pre-modern craftsmanship, based on hands-on experience, and our modern science-based technology. From my first visit to the Archeon archaeological theme park in the Netherlands I remember in particular the pre-history area, then directly behind the main building. I recall the unearthly silence on entering it, the sensation of being completely out of earshot of radio stations, muzak, the beeping of reversing lorries or the neighbour's power drill, and the relaxing sensation of not being semi-consciously occupied with blotting them out. And therefore: silence, pre-modern silence.

Loud noise, obviously, has always been with us; in ancient Kaifeng or Rome the racket must have been considerable, and probably round-the-clock. But it was easy to escape from it; you only had to walk out of the city gate. More importantly, pre-modern noise was not intended to numb the brain through automated drumming or the electronic amplification of mindless babbling. Our modern urban world is not only full of inescapable cacophony, but for the first time in history the (in principle) pleasing sound that we call music has become part and parcel of that noisy background. How could our daily experience of sound change so radically? What or more importantly *who* was behind it? Who brought about the disruption of a centuries-old pattern, which all over the globe may have varied somewhat from place to place, but whose underlying features were the same everywhere?

Well, Hertz and Marconi brought it about. Though not alone, of course. The ingenious physicist and the clever engineer did not compose a single note of music and never deafened the neighbourhood with quadraphonic speakers blaring from their cars. They would have been staggered to find that any such thing could have resulted from the discovery made by the former and the invention of the latter. Yet, for all their stunned amazement, they would have had to concede that without the theoretical prediction of the radio wave and its concrete application in the 'wireless telegraph', modern electronic sound could not have happened as it did, or for that matter at all. At best they might point their finger at later scientists and engineers who built on their novel and, as such, still elementary insights and attainments. Or alternatively they might go back in time and single out Maxwell as the great theoretician of the electro-magnetic continuum, who in his turn might refer back to Faraday as the great investigator of electro-magnetic effects, who would then point at Newton as his role model, who in his turn would refer us back to Galileo as the great trail-blazer of an approach to nature that really holds water. Indeed, we can find such references in their collected works and letters. True, Galileo used to invoke at times the 'divine' Archimedes, but he was well aware that his own way of examining natural phenomena was not only in the right direction but also had no real precedent.

That is how we arrive at the years around 1600, which is when 'the re-creation of the world' (to use the original, Dutch title of this book) began. This beginning consisted first and foremost of new ways of thinking. For centuries Greeks but also Chinese, Europeans but also Arabs, monks but also laymen, lone thinkers but also philosophical schools, had reflected with great acumen and perseverance on how the natural world hangs together. But traditional thinking, though often very ingenious, was, in retrospect, most often misdirected, and between the 1600s and the 1640s Galileo, Kepler, Descartes, Bacon and numerous others gave a decisive new twist to it. It was not just a theoretical twist but, inextricably bound up with it, a practical one as well. The kind of reasoning associated with the embryonic modern science emerging at that time has been called 'hands-on thinking' (literally 'thinking with the hands'). For the first time ever, room was created for testing systematically whether assertions about nature stood up to reality. All over the seventeenth century hands-on thinkers began to explore procedures and practices for checking whether a plausible-looking assertion was anything more than just a plausible assertion. In this book I seek to analyse how all this happened in the way it did and to explain what made it possible.

Expert historians of science have provided numerous accounts of the seemingly miraculous range of theoretical and practical breakthroughs from Galileo to Newton. My own introduction to the genre was E. J. Dijksterhuis' *The Mechanization of the World Picture: From Pythagoras to Newton*, published in Dutch sixty-five years ago but still in many ways an inspiring account. I myself drew up a comparative and critical inventory of all those dozens and dozens of interpretations and explanations in *The Scientific Revolution: A Historiographical Inquiry* (1994). But no systematic, source-based effort has yet been undertaken to explain why the decisive move towards modern science happened in, of all places, Europe, that latecomer among the great traditional civilisations. Why not in China, why not in the Islamic world, in both of which the knowledge of nature was at times pursued in quite advanced ways? There are enough clichés in circulation, and plenty of glib answers, but until recently there had been no sustained in-depth historical comparison of the pursuit of nature-knowledge in these three civilisations. In my *How Modern Science Came into the World: Four Civilizations, One Seventeenth-Century Breakthrough* (2010), which is just such a comparative study, I set out my research findings in full, complete with an account of the book's conceptual underpinnings and sources and references to the literature. In the present volume, which contains the same argument in shortened form, I am not primarily addressing my fellow historians of science, but a wider readership. I set myself up here as the reader's authoritative and reliable guide – a posture that I dare to assume only in view of the existence of that larger book, in which I instead invite the reader to join me on an adventurous voyage of discovery.

The wider readership of the present volume needs no special knowledge to follow the argument developed in it. I set forth in words, not formulas, the mathematical issues that arise from time to time. Far more important than the knowledge that the reader may or may not possess is a willingness to set that knowledge aside for the time being. I began this chapter with the contrast between the Old World and the New. We can only conjure up life in the Old World by looking around us and then, one by one, eliminating things. Let's pull out that plug, get rid of the gas oven, ditch your mobile phone. Hey, what's that plastic bin bag doing here? That bike in the wooden shed can go too. The shed can stay, though. In the same vein I am asking you, the reader, to delete several modern concepts from your brain. Say good-bye to evolutionary theory, abandon the law of universal gravitation, jettison the table of chemical elements. I must even beg you to suspend for now your own notions of how our knowledge of nature originated and how that knowledge advances over time. If you've read Kuhn, don't immediately see paradigm shifts everywhere; we'll find out in the course of the book whether they are of any use to us. If you prefer Popper, try abandoning your trusted criterion of falsifiability until the final pages. And if you are tacitly assuming that science in the past must really have been about the same as that of today, only much simpler and with weird errors that the great heroes of past science managed over time to weed out one by one, then please clear a space in your brain for a situation where there is no such thing as science at all. The natural world now lies before us unexplored and undiscovered; how can we, through thought and observation, come to grips with it?

1   To begin at the beginning: nature-knowledge in Greece and China

The natural world around us looks both impressive and mysterious. In the past, controlling it in times of drought or plague required magical incantations, while real understanding came via the world of the gods. Take the *Iliad* or the *Odyssey*: the angry voice of Zeus (Jupiter) is heard in a thunderstorm; volcanic eruptions and earthquakes are caused by Hephaestus (Vulcan) hammering on his anvil; if rain should fall while the Sun is shining, Iris hurries to place a rainbow in the heavens. In the pantheons of other civilisations it was much the same if with different names. But such explanations still left open the possibility of penetrating more deeply into specific phenomena. The Babylonians, for instance, produced strikingly accurate predictions of the positions of the Moon, stars and planets by systematically tracking their movements through the night sky. The Polynesians, by sophisticated observation of subtle changes in cloud formations and bird flight, were able to navigate their canoes accurately over hundreds of miles of ocean.

Among the civilisations that developed such specialised nature-knowledge, two took a further decisive step. They were the Greeks in the sixth century BC and the Chinese at about the same time. Both ceased appealing to explanations of the Zeus/Iris type, and came up with a very different picture of the natural world. They did not abandon their belief in gods and the spirit world, but they no longer attributed the myriad of natural events to divine action. Instead, they posed certain principles of natural order and established certain explanatory schemata that enabled them to understand and chart the whole of the natural world from a few fundamental points of view.

There are of course many ways of doing this, and just as one can choose between eating with a knife and fork or with chopsticks, or writing with letters or characters, so one can choose between different ways of approaching natural phenomena and breaking them down into manageable portions. Accordingly, how the Greeks chose to approach and order the natural world turned out to be very different from how the Chinese did essentially the same thing. The Chinese approach relied primarily on observation and focused on practical use. In the second century, Zhang Heng attempted to detect regularity in the occurrence of earthquakes in order to find a way of predicting them. His observation-based research was conducted against the background of a coherent world-view that had taken shape gradually and had allocated to each phenomenon its specific place. The Greek approach, by contrast, was not 'bottom-up' like the Chinese, but 'top-down' – i.e. generalisation preceded the collection of data, and the observed facts were fitted into an intellectual construction. The empirical element was minimal; the thinking was abstract and theoretical. And whereas in China, after the unification of the empire under one emperor in 221 BC, a synthesis emerged in which one approach and one world-view largely prevailed, in Greek thought a permanent division developed. In Athens abstraction and theorising took the form of philosophy; in Alexandria of mathematics. For instance, Athenian philosophers explained in broad outline the structure of the cosmos from the Earth up to the outer sphere of the stars, whereas in the Greek colony of Alexandria mathematicians used models to calculate the planets' trajectories through the heavens.

The existence of that dichotomy between the Athenian and Alexandrian approaches is a key element in this book. Without an insight into its whys and wherefores, it is hardly possible to explain the much later emergence of modern science. Taking that as our starting point, we shall first consider the Athenian mode of nature-knowledge and then the Alexandrian, before charting how they differed and how deeply rooted their differences were.

IN ATHENS

Philosophy is all things to all people. It provides solace, worldly wisdom, mental discipline, ideas on good statesmanship and advice on how to deal responsibly with our fellow human beings. Each of the four schools of philosophy that were founded in Athens more than two thousand years ago had answers to such questions. Furthermore, they each had a fully worked-out conception of the natural world and claimed to have an understanding of nothing less than its very essence. Whether you visited the Academy which Plato founded, the Lyceum of his pupil Aristotle, the Stoa's colonnade or the garden of Epicurus, you would always find someone to explain to you the ultimate source of natural phenomena. Of course, the views of each of these schools differed but there was nevertheless one thing that they all had in common: they each had their own explanation for the same problem, the problem of change.

How can change be a problem? After all, things are changing around us all the time. The branch of a tree blows off and falls to the ground, water evaporates in the Sun's heat, a volcano spews out lava, a child becomes an adult and eventually withers away. Surely the search for nature-knowledge should focus on detecting regularity within this ceaseless change? Had not a very early Greek thinker, Heraclitus, expressed this in his celebrated *panta rhei* (everything is in flux ... nothing endures but change)?

But change was turned into a problem by that most contrary of the early philosophers known as the 'pre-Socratics', Parmenides. In fifty-odd lines of didactic verse, he declared that change is a delusion. There is only Being and Non-Being and no intermediate or transitional form can be conceived of without inner contradiction. For if that into which something changes did not originally exist, where could it possibly have come from? Or it was there from the beginning, in which case there is no change at all since everything has remained as it always was:

> How might what is then perish? How might it come into being?
> For if it came into being it is not, nor is it if it is ever going to be.
> Thus generation is quenched and perishing unheard of.

In short, it is unthinkable that between Being and Non-Being there exists a category of Becoming. If we believe that we observe change all the time in our daily lives, then our observation is at fault and we must conclude that our senses fail to provide us with reliable information about the real world. What is at one and the same time impressive and off-putting about early Greek thought is that cold-blooded readiness to draw such a conclusion. And as so often happens in the history of ideas, inspired nerve brings its rewards. There are scholars who believe that those fifty-two rather obscure lines of verse by Parmenides have determined the direction of the Greek philosophical tradition ever since.

Sensory perception, so peremptorily cast aside by Parmenides, was soon rehabilitated, although it was not by pretending that nothing had happened. The *paradox* of Parmenides, that in spite of appearances Becoming is impossible, is creatively converted into the *problem* of Parmenides: how can we recognise the validity of the paradox and at the same time render it harmless through equally rigorous reasoning? Can we rescue 'Becoming' from its logical difficulties and make it comprehensible after all?

The fundamental principles adopted by the four philosophical schools that were established in Athens in the period after the pre-Socratics all had an answer to the problem of Parmenides.

Plato goes along with him the furthest. He makes a distinction between the imperfect world that we observe with our senses and a perfect world of Ideal Forms of which the objects of sense are merely a poor reflection. There are pine trees and oaks and palm trees to which all kinds of things may happen, but what really matters is the unchanging Idea of the Tree, the ideal tree from which all those pines, oaks and palms take their specific form. Knowledge of nature, of human beings and of human society is always concerned with their ideal forms. Plato's prime concern was the constitution of the state, which should embody as closely as possible the Idea of Justice. But in one of his dialogues, the *Timaeus*, he provides an insight into his view of the natural world. He recounts a creation myth in which nature is designed by a wise architect modelled on certain regular geometrical figures. The significance of mathematics for Plato lies in the way it highlights the distinction between the world of daily experience and the ideal world which can only be imagined, not observed. If you draw a triangle in the sand and measure its angles they will add up to, say, 175 degrees, or if you have drawn and measured it very carefully, even close to 180 degrees. But you will never attain in this way the exact 180 degrees, 0 minutes, 0 seconds that is the proven outcome. The addition is exact only for the triangle that we hold in our minds, its purity untouched by our senses. That is how learning mathematics can prepare us for the highest form of knowledge, which is a mental vision of the Ideal Forms in the Realm of Ideas.

Plato thus resolves the paradox of Parmenides by recognising the existence of change but declaring it to be of secondary importance; the reality that truly matters remains unaffected. His pupil Aristotle, in contrast, accepts change as a fully fledged category in its own right. His innovation is that he distinguishes between two kinds of Being: Potential and Actual. Change can now be understood as the growth of what something carries inside itself as a possibility into a state of actual reality. An example is an acorn (only potentially an oak) growing into a fully formed oak tree. Change is thus the realisation of an end that from the start lies hidden within that which changes. The essence, the ideal form, does not – as Plato taught – reside in another world, but within things themselves, and all things are engaged all the time, with more or less success, in achieving their ideal form.

In Epicurus' atomist doctrine, Parmenides' paradox is resolved in yet another way. The One immutable Being is broken up into an infinite number of minuscule fragments of Being called 'atoms' (from the Greek word for indivisible). They are infinitesimally small, indivisible particles of matter which move unceasingly through otherwise empty space. All change in the world comes about through the never-ending reconfiguration of these atoms, which combine to create heavenly bodies or boulders or living creatures but which eventually separate out again to merge into new combinations.

Atomism saw the world as a compound of separate units. For the Stoa, by contrast, the central theme is the close inter-relationship between things: not their physical separateness but rather their unbreakable continuity. Here change takes the form of ceaselessly shifting interactions within the *pneuma*, an active inter-mediate substance which permeates the entire cosmos. The Stoics conceived of the *pneuma* as an extremely subtle rarefied mixture of fire and air which is both material and spiritual, and which holds everything together. And just as a spider detects a fly by the slightest movement on its web, so the fluctuations of tension and release in the *pneuma* reflect all natural activity. Another well-worn analogy is that of the pebble thrown into a calm pond. The ripples propagate the initial tension across the entire surface of the water.

These four views of change and the resultant conceptions of nature differ quite basically from each other, and the differences gave rise to much debate – Greeks were keen debaters. A certain supporter of Aristotle challenged the Stoics to explain how so tenuous a sub-stance as *pneuma* could ever hold together objects that are so much more solid and heavy. The Stoic response was that the *pneuma* derived its cohesive powers from the elasticity of Air and the inces-sant action of Fire, its two constituent parts. This kind of exchange was remarkably sterile – neither side was prepared to abandon its fundamental tenets. Underlying the criticism of *pneuma* lay the fact that Aristotle had introduced a similarly rarefied substance into his philosophy of nature, but one that remained celestial and exercised no autonomous function. *Pneuma* could not have fitted into Aristotle's system even if he had wanted it to. The basic principles and respective points of departure of the two philosophies were mutually exclusive. Because each of the Athenian schools insisted that their basic prin-ciples were so immediately self-evident that they left no room for doubt, agreement between them was quite out of the question.

So they did have something in common after all – the way knowledge was structured. Each of the four schools started out with a small number of first principles which were purportedly the distinguishing marks of the real world around us. Each set of first principles was taken to be so self-evidently certain that their truth could not be questioned, let alone disproved. There is no limit to the causal effect of this approach, for every observable fact fits seamlessly into a whole that has already been determined by the first principles. This is not to deny that each Athenian school had some favourite observations. The atomists, for instance, would point out the atom-like particles of dust revealed by a ray of sunshine entering a darkened room. Or they would cite the wearing away of a stone step: an undetectable layer of atoms is removed by every passing footstep until in the end there is no step left. For the Stoa a comparable role is played by the stone that creates ripples in the surface of a pond, an observation that simultaneously illustrates the doctrine and is explained by it.

Not only did each Athenian school create its own view of the world, but it also did so in a specific manner for which I shall reserve the term *natural philosophy*. We shall encounter other world-views, both in China and in Renaissance Europe, which were not natural philosophies as defined here, but much looser mental constructs. *World-view* will always signify here a broad conception of how natural phenomena hang together. I shall employ the concept of natural philosophy much more narrowly and only if its world-view is furnished with that specifically Athenian structure of self-evident first principles that characterise the world in its totality.

The indubitable certainty of each separate set of first principles suggests that something unusual was afoot. If there had been just one system of first principles you might perhaps persist in maintaining that the world must be as laid down by those principles. But the confrontation between as many as four sets of such principles made it much more difficult to argue that any one of them could have a monopoly of the truth. And it was at this weakness that a fifth Athenian school directed its criticism, not in the guise of yet another philosophy but rather to flesh out an anti-philosophy. This was the Sceptical School, whose founder, Pyrrho, concluded from the existence of four philosophical systems, each claiming complete certainty, that human beings could never enjoy certain knowledge. Both our intellect and our senses can deceive us in all kinds of ways, from the dreams that confuse our minds to the colour blindness that distorts our observation. What we believe to be knowledge is only seemingly so; in reality we can know nothing with any certainty. This standpoint quickly provoked the objection that Pyrrho was in fact claiming certainty for at least one assertion, namely that we cannot know anything with certainty. Consequently, Scepticism shifted towards a position known as 'the suspension of judgement'. We cannot with any certainty make any generalisations about things that we cannot immediately observe. 'I can see and feel that it is raining' is about as far as one can go. What a sad outcome to the fearless philosophical adventure that the pre-Socratics had initiated five centuries earlier!

IN ALEXANDRIA

Greek thought might never have developed any further if, far away from Athens, a different approach had not emerged. After Alexander the Great had conquered the eastern Mediterranean, the city of Alexandria which he founded in Egypt came to rival Athens as an intellectual centre. As in Athens, a number of pre-Socratic ideas were taken up and developed further. Primarily involved was mathematics, a discipline that was not a Greek invention. Surviving hieroglyphs and clay tablets show that, centuries earlier, fairly advanced mathematical understanding had already developed. Egyptians could calculate the volume of a truncated pyramid with a square as its base. Babylonians could sum arithmetic series and solve quadratic equations. The big difference is that Greek mathematicians came up with the idea and the practice of mathematical proof. The Babylonians were well aware that in a right-angled triangle the square of the hypotenuse is equal to the sum of the squares of the other two sides, but they formulated it as an operational rule with examples that a land surveyor would find useful. In the fifth century BC, Pythagoras turned it into an abstract theorem, omitted its practical applications and added mathematical proof. Plato found this kind of abstract mathematics very useful as an introduction to his doctrine of Ideas. In the *Timaeus* he even turned certain triangular figures into the foundation stones for his conception of how the world was created. But, as Alexandrian mathematics rapidly developed further, its connection with the natural world came to diverge quite widely from Plato's. Again Pythagoras provides a good example.

He is said to have discovered *mathematical* regularity in a phenomenon that all of us can observe with our *senses*. Two random musical notes do not normally sound well together: they appear to clash, they are 'dissonant'. But now and then among the many dissonances a musical interval makes its appearance where the two distinct notes seem to melt together and sound pleasing to the ear; they are 'consonant'. These consonant intervals can be produced systematically by sounding a string, first over its full length, then half its length, that is, in a ratio of 1 to 2; similarly in a ratio of string lengths of 2 to 3, and finally of 3 to 4. This produces respectively the octave, the fifth and the fourth. This is how Pythagoras could conclude that those rare consonant musical intervals correspond exactly to the ratios of the first few integers.

The discovery of this unexpected natural regularity led him to reduce at one inspired stroke the whole wide world to number. According to him and his followers, the cosmos itself is structured according to the ratios that exist between harmonious sounds, the so-called harmony of the spheres.

These harmonic intervals provide the sonic material with which to build up musical scales. Another great Alexandrian mathematician, Euclid, also worked on these consonant ratios, all the while steering clear of any speculation about the structure of the world: from his point of view, the harmony of the spheres was more of a plaything for philosophers. Euclid confined himself to a specialised treatment, with rigorous theorems and proofs, of the numerical ratios involved in the successive divisions of a string.

That specialised approach remained a characteristic of the mathematically based variant of nature-knowledge which flourished in and around Alexandria from about 300 BC onwards. Among its practitioners were the greatest mathematicians of ancient Greece, who also took 'pure' mathematics to great heights; not only Euclid, but also Archimedes, Apollonius, Aristarchus and (centuries later) Ptolemy. Five natural phenomena in particular lent themselves to the kind of abstract mathematical treatment that was applied to musical consonance. They were light rays, equilibrium states of solid and floating bodies, and the observed positions of the wandering heavenly bodies: the Sun, the Moon and the planets. But what was actually 'mathematical' about these widely differing phenomena?

It is most obvious with light because it travels in a straight line, the simplest geometrical figure. By drawing lines to represent light rays, Ptolemy was able to derive some basic rules for reflected light (angle of incidence equals angle of reflection) and for the refraction of light when entering water (even though his mathematical rule later turned out to be no more than approximately valid).

The law of the lever, which states the conditions under which a beam with weights hanging from it remains in equilibrium, is less obviously mathematical. It requires a process of artificial abstraction, free from the material world. The beam is represented by a straight line; the weights are not actually suspended from it but just considered to be associated with it.

The same Archimedes who proved the law of the lever also worked out rules governing the floating or sinking of objects in water: rules that boil down to distinguishing the specific gravity of different materials. There again the capricious shapes of objects in the real world are abstracted into geometrical figures.

Finally, the positions of the heavenly bodies in the firmament. If you follow them from night to night, the stars appear to move through the heavens according to a fixed pattern, whereas the Moon and the planets follow a pattern which is much less obvious.

So it is with the Sun. The seasons do not all last the same length of time: from the spring equinox through summer to the autumn equinox takes nine days longer than from the autumn equinox through winter back to the spring equinox. One irregularity in the behaviour of the planets is that their brightness varies over time, suggesting that their distance from the Earth is not constant. Also when their paths are observed from night to night relative to the signs of the zodiac these paths prove not to be rectilinear – now and then the planets turn back on their tracks (this is known as their retrograde motion). Even so, Alexandrian mathematicians wanted to be able to predict exactly when a planet would be at a specific point in the heavens. In doing so they held firmly to three principles. The Earth is motionless at the centre of the universe and the heavenly bodies that revolve around it do so in *circles* traversed *uniformly* (with a constant speed). However, the apparent paths which they seem to follow with their irregular about-turns are far from circular. This gave rise to the construction of complicated models, first completed in a book by Ptolemy (known by the title of its later Arabic translation as the *Almagest*). Characteristic of these models is that they consist of complex combinations of circles which could never exist in celestial reality. They are fictional, with one great advantage: if you entered the current position of a planet into the model, its prediction of future positions appeared to match accurately the best observations then made.

Each of these five domains of mathematical nature-knowledge is therefore characterised by a high level of abstraction and their links with the actual natural world are extremely tenuous. The only serious exception is Eratosthenes' well-known determination of the Earth's circumference. For the rest, mathematical theory and natural reality remain almost entirely separate entities. A real, material lever does not obey the law of the lever all that closely even though its deviation from strict mathematical exactness becomes smaller the closer it approximates the simplicity of a straight line. Straight lines are drawn to demonstrate how rays of light are reflected and refracted, but what in refraction actually happens at the interface between air and water is left open. And in discussions of consonant intervals the vibrations of the string which produce the sounds that make up the intervals are left out of consideration entirely. Musical theory cultivated the mathematical way remains confined to abstract manipulations of the ratios that you get when you make the entire string vibrate, then its half, then one-third and so on.

The last great Alexandrian, Ptolemy, was well aware of this high level of abstraction, and in his writings on planetary trajectories, on consonance and on the reflection and refraction of light he made sustained efforts to strengthen in each case the connection with the real world. He did his best to extend the circle combinations in the *Almagest* to three-dimensional schemata. In his textbook on astrology he linked up the planets and the angles from which they are observed with the climate on Earth and with human fate. He also tried to establish a synthesis between the Pythagorean account of consonance and a treatise written by a pupil of Aristotle's that privileged melody over harmony. To his treatment of the paths of light rays in reflection and in refraction Ptolemy added a discussion of the workings of the eye. Looking back, the modern scientist in us may disparage or even scorn Ptolemy's obvious failure to build truly effective bridges between his mathematical models and physical reality. But the historian in us finds it more rewarding to see his failure as a measure of the huge conceptual problems that had yet to be overcome. Only in retrospect can it appear that there is a direct line between an almost fully abstract form of mathematical science and one that is much more closely linked to the reality of natural phenomena. In this book I aim to show how certain pioneering scholars have managed to get there, partly through a range of coincidences but also through an orderly series of events in which clear-cut patterns can be discerned. One of those patterns lies in the contrast between the two varieties of Greek nature-knowledge to which we have just been introduced.

ATHENS AND ALEXANDRIA: TWO MODES
OF NATURE-KNOWLEDGE COMPARED

Mathematical nature-knowledge in its Alexandrian guise, then, was barely connected with the real world. That is one of the most important differences from Athenian nature-knowledge, which was firmly oriented towards the world of everyday reality, even if considered from a particular viewpoint. This applied even to Plato, who was so concerned to transcend daily reality. A further difference is that in Athens, unlike Alexandria, the aim was to *explain* phenomena, and to do so by giving them their rightful place in an overarching whole. Natural philosophers set themselves the task of explaining the world from a few basic principles that were not susceptible to any kind of doubt. All observable phenomena were understood in terms of those basic principles and served at the same time to illustrate them, with each philosopher selecting those which fitted best. A unified, all-embracing explanation, qualitative rather than quantitative, derived from forever-valid first principles and laid out in ample discursive treatment: that is what the natural philosopher aimed for. Not that *natural* philosophy was his sole concern. His views on the nature of the universe were closely bound up with other core philosophical questions such as how to organise the state, how to live a virtuous life and how to compose and present a well-reasoned argument.

Mathematical nature-knowledge, on the other hand, is self-contained. The validity of a particular theorem does not stand or fall with the truth of some adjacent insight. First principles of the type that, for an 'Athenian', lie at the heart of everything leave an 'Alexandrian' cold. He does not want to explain, but to describe and prove, and he does not do it qualitatively in eloquent prose, but in mathematical units – figures and numbers. Observed phenomena are not used as illustrations but as starting points for mathematical analysis, which from then on is almost entirely abstract. The five phenomena (consonance, light, planetary trajectories and the two states of equilibrium) are investigated separately. There is no search for interconnections, let alone for an overarching unity. The one thing that they have in common is that they lend themselves to mathematical analysis. But while any educated person can take part in philosophical debate, mathematical nature-knowledge requires a considerable degree of expertise. The few who do possess it tend not to confine themselves to one specialist area but as a rule excel in several fields. Euclid, for instance, not only produced his celebrated, systematic survey of virtually the whole of Greek mathematics, but also wrote treatises on musical consonance and on light. Archimedes became the authority on both types of equilibrium states, while Ptolemy's work constituted the pinnacle of 'Alexandrian' thought in the other three domains.

In sum, then, 'Athens' and 'Alexandria' were fundamentally different from each other in more than just their contents. They were two different *modes of nature-knowledge*. That is not a recognised technical term but I have introduced it as a viable alternative to speaking of 'natural science'. For present-day readers it is very hard to avoid loading the kind of nature-knowledge that we shall be dealing with in this and later chapters with concepts and practices that have in due time come to mark *modern* science. The Greeks conceived of the atom, we use it still, and that makes it easy to ignore that, apart from the word itself, the doctrine of ancient atomism has hardly anything in common with the modern theory, and was in any case embedded in a very different conceptual framework. In addition, the modern eye almost invites us to judge much of what we have surveyed in this chapter as seriously deficient. To us, it is so obvious to associate musical intervals with the vibrations of the string that produces them; what then prevented the Greeks from Pythagoras to Ptolemy from taking that seemingly simple step? For the historian this is not a useful question. More to the point is to observe that Greek thinking about consonance, or about any other subject, took place within one of two distinct frameworks, neither of which was that of modern science (even though the Alexandrian in particular had certain important traits in common with it) but *which had their own characteristics and their own potential for further development.*

I use the concept 'mode of nature-knowledge' as my fundamental unit of historical analysis throughout this book. Each mode of nature-knowledge is, as it were, a self-contained package, a coherent approach to natural phenomena, which can differ from others in many ways. They can differ in their scope: all-embracing like the Athenian or partial as in Alexandria. They can differ in the degree to which they are oriented towards day-to-day reality, in other words in their level of abstraction. They can differ in the way in which knowledge is acquired, whether primarily through the intellect or via the senses. They can differ in associated practices, such as observation, experiment or the use of instruments. Even their ultimate goals can be different: knowledge might be sought as a goal in itself, or in the hope of achieving some practical objective (e.g. in navigation, or to reduce manual labour). Finally, there may or may not be any interaction between one mode of nature-knowledge and another, as with the striking *lack* of interaction between the Athenian and Alexandrian modes of nature-knowledge that we shall shortly be considering.

Of special importance in all this is what I further call the 'knowledge structure' of a particular mode of nature-knowledge. Was the sought-for knowledge intended to explain or to describe? If the latter, would the description be in words or in measure and number? Are the observational data treated as independent units of knowledge or as part of a pre-established scheme? If the latter, do they serve to illustrate and confirm the scheme or to assess critically whether the scheme is truly sound? And what about the orientation in time? Do those who practise a particular mode of nature-knowledge believe they are restoring the lost perfection of an earlier age, or do they hope to construct a new all-embracing system, or do they see themselves as contributing to a future that in principle is open-ended?

These distinct modes of nature-knowledge are not necessarily set in stone. Circumstances may appear that lead to their being subjected to transformation. Indeed, here is where the central problem addressed in this book will eventually be resolved. Transformation can occur as limited expansion and enrichment along pre-established lines, yet such expansion may also (but need not) morph into truly revolutionary transformation, in spectacular ways which we shall encounter more than once in seventeenth-century Europe.

'Modes of nature-knowledge' is not a term that our protagonists ever used. But neither did they call themselves natural scientists; the expression 'science' for more than a single branch of research dates from the nineteenth century. A multitude of other expressions were in use which do not lend themselves to establishing a common denominator. So it is best to select the terms we shall be employing here independent of contemporary usage. In so doing we must take care to define them with sufficient precision, and make every effort to avoid the pitfalls of verbal confusion. I shall therefore refer to the Alexandrian mode of nature-knowledge as 'abstract mathematical nature-knowledge', or as 'Alexandria' for short. For the Athenian mode I shall continue to use the term 'natural philosophy', or 'Athens' when we need to emphasise the contrast with 'Alexandria'.

Should we then really speak of 'Athens' and 'Alexandria' as two separate modes of nature-knowledge in the sense that we have now given to the concept? The list of contrasts that I have just drawn up shows that we should certainly do so: the abstract-mathematical approach differed in many striking respects from that of natural philosophy. In fact, they were so fundamentally different that virtually no exchanges between them took place. This has partly to do with their origins and the large geographical distance separating them. But more was involved. 'Alexandrian' texts show that their authors were well aware of some elements of Athenian natural philosophy. Ptolemy on occasion would have recourse to them to plug gaps in an otherwise strictly mathematical argument. But such incidental dipping into the available supply of natural-philosophical concepts was pure opportunism. Ptolemy thought that natural philosophy, which he curtly dismissed as 'guesswork', could in no way enjoy the certainty provided by mathematical deduction.

There are nevertheless a few points of contact at the level of actual content that might have been exploited but (characteristically) were not. For instance, the mathematical account of consonance might have been linked without much difficulty to the Stoic idea of sound being propagated by wave-like movements of the air. That it did not happen had to do with those principal differences separating the two modes of nature-knowledge. But there is also such a thing as the stability of an established tradition. Certainly in the Old World things could be expected as a rule to stay the same; innovation was the exception. This was surely the case with such intricate intellectual constructions as the two Greek modes of nature-knowledge. We shall see how the Athenian and Alexandrian modes were enriched and even transformed independently of each other until in the mid seventeenth century Christiaan Huygens brought two still quite distinct modes of nature-knowledge together in fruitful interaction for the first time.

In the Greek world, where these two modes of nature-knowledge first developed, genuine interaction was out of the question, although inevitably there was some overlap. The prime example here is the universally shared belief in, or more accurately the tacit because entirely self-evident acceptance of, a stationary Earth at the centre of the universe. Throughout Greek antiquity there was only one person, Aristarchus, who unequivocally disagreed. But, so far as we know, he never developed his thoughts into a full-blown mathematically based planetary theory as Copernicus was to do fifteen centuries later. Certainly the odds were clearly against it. After all, we can't *feel* the Earth turning, can we? Surely we can all *see* the Sun rising in the east and setting in the west? The intellectual elite could be aware of the spherical shape of the Earth from the time of Parmenides, but that it should spin on its axis once every day or revolve around the Sun each year was quite another matter. There was no observable evidence for it and in any case it defied common sense. Ptolemy well knew that a rotating earth would simplify his mathematical models on a number of points but he also considered the objections to it were too strong. And so, at least on this point, Alexandrian mathematical astronomers such as Ptolemy could agree with an Athenian natural philosopher such as Aristotle who was the most specific among his colleagues about the place of the Earth in the cosmos. Except that for Aristotle a stationary Earth at the centre of the universe was more than just a factual given. Not only *was* it so; it *could not be otherwise.*

Indeed, the way in which Aristotle designed his cosmos offers another clear-cut example of what the Athenian mode of nature-knowledge looked like. Aristotle solved the Problem of Parmenides by conceiving of change as the transition of potential Being to actual Being. Change is therefore the attainment of an end that is an essential, inherent characteristic of that which changes. Change occurs in four different ways: as generation and corruption (out of an acorn grows an oak tree); as qualitative change (green leaves turn brown in the autumn); as quantitative change (fewer leaves are left in the autumn); and as change of place (leaves fall in the autumn). Even in change of place, or movement, the moving object seeks to attain an end. What end that is depends on what primeval matter the object is made of. If it consists mainly of earth, which is heavy, or water (less heavy primeval stuff), then the object will move in a straight line to the centre of the universe. If it consists mainly of air (light primeval matter) or of fire, which is even lighter, then it will move in a straight line *away from* the centre of the universe. Finally, if it is made of immaterial 'ether' it will move in a circular path *around* the centre of the universe. In a natural order that is fully realised, when all ends have been attained, there will have been formed around the centre of the universe a sphere of earth, around that a spherical shell of water, then a shell of air and then one of fire. Around that the ethereal heavenly bodies move forever along their circular paths. The reader will have already recognised it. Earth at the centre with the seas and oceans surrounding it? The atmosphere around that? The Moon and Sun, the stars and planets, revolving daily around it all? It is our own world that Aristotle was able to derive from the basic principles of his natural philosophy.

Only, it is an 'Athenian' world that is being explained here by means of careful deduction, not the 'Alexandrian'. They agree on a stationary Earth at the centre of the universe, yet that is where agreement ends. Aristotle explains in words the broad outline of a universe in which he assumes that every heavenly body follows a simple circular path around the Earth. Ptolemy describes in a mathematical model built up of a large number of circle-combinations the paths of the planets in their full complexity (at least to the extent that they were known at the time), and says nothing about the rest. Their aims are different, as is, to a certain extent, the outcome. Little or no need is felt to unify the Athenian and the Alexandrian points of view. At most, in an effort to enhance the reality content of his abstract mathematical derivations, Ptolemy makes use of an occasional snippet of natural philosophy.

We have talked about Ptolemy as if he were more or less a contemporary of the other great 'Alexandrians' such as Euclid, Archimedes, Aristarchus or Apollonius. But he was not; he did not live during the great flowering period, but several centuries later in the second century AD. He was (as far as we know) a solitary person of genius living in an age long after the decline of Greek thinking about nature had occurred. That decline now calls for our attention.

DECLINE: AN ESTABLISHED PATTERN

First the basic facts. Around 150 BC the pioneering work in natural philosophy had been completed and much the same is true of mathematical nature-knowledge. Other cultural activities continued to flourish while some had not even begun; the decline and fall of the ancient world still lay far in the future. So we are not facing a universal process of decline, but only the particular circumstance that the Golden Age of Greek thinking about nature comes to a fairly abrupt halt in the mid-second century BC. That Golden Age was the culmination of a creative period that had lasted for a good two and a half centuries from the 'overture' of pre-Socratic thinking and the first schools set up by Plato and Aristotle. The explosion of achievement that followed (from about 350 BC) was the first of a series of Golden Ages of nature-knowledge that we shall encounter.

What is meant here by 'Golden Age'? I define it as a remarkably dense concentration of great creative talent. In the case of Greece this involves, in alphabetical order, Apollonius (researcher into conic sections), Archimedes, Aristarchus, Chrysippus (inventor of Stoic *pneuma*), Epicurus (founder of the atomist school), Euclid, Hipparchus (pioneer of mathematical planetary theory) and Pyrrho. And they were not the only ones; they were just the greatest and most original. At a lower level they were surrounded by many others, in far greater numbers than the half dozen or so individuals per generation who actively pursued nature-knowledge before the Golden Age. But when Hipparchus dies in about 150 BC the explosion comes to an end. He is the last, and within a single generation we find at best second-rate adherents bombarding each other with the arguments of their respective teachers. But though it was an age more interested in going over old ground or pursuing entirely different subjects, there was always the possibility of an intellectual giant appearing who would calmly pick up and continue where others had left off centuries before. Ptolemy is the best example of this intellectual 'afterburn effect', although right up until the fifth century there were a few more, both Alexandrian and Athenian. The appearance of these incidental 'afterburners' is not in itself an argument against the observation that what followed upon the Golden Age was a steep and major decline. Indeed, it is characteristic of the rise and fall of nature-knowledge in the pre-modern world. First the overture, then a flowering which culminates in a Golden Age, and finally a steep decline during which, now and then, an exceptional individual finds the way back to the top. History never repeats itself exactly; even so we shall come across essentially the same pattern in Islamic civilisation as well as in medieval and Renaissance Europe.

Why then decline? What caused it? Historians have given much thought to the problem but this is really not the right way to pose the question. For the answer is: what else would you expect? We are back in the Old World, far away from modern science and the two sturdy pillars on which it rests in modern society. In our own time the continuity of the scientific enterprise is due to two extraordinary factors. Modern scientific research is propelled in part by an inner dynamic due to which the frontiers of knowledge keep being pushed back. Also, much of that knowledge, in ongoing interaction with a largely science-based technology, has proved capable of increasing prosperity and in many respects improving public wellbeing year in and year out, all of which lies at the heart of the array of contrasts with which this book begins. Nowhere in the Old World could any form of nature-knowledge even at its height rest upon either of these twin pillars of the modern scientific enterprise. There was no inbuilt continuity; it is indeed the task of this book to explain how any mode of nature-knowledge was able to emerge that was *not* bound to decay within a century or so. In the Greek world, it is not decline but a *failure* to decline which would urgently demand an explanation. At most, it makes sense to ask why decline set in when it did, in the middle of the second century BC, and also to enquire about the specific nature of that decline. Did creativity just cease, or was it diverted into other channels?

The first question – 'Why just then?' – cannot be answered with any certainty. But we may note that two elements that in later civilisations would co-determine when decline actually occurred did not play a role in ancient Greece, namely large-scale destructive invasions and the suspicion of sacrilege. In societies with a Sacred Book, as we shall see, all kinds of objections are made against nature-knowledge. But in the ancient world with its many gods it was always possible to make fun of natural philosophy, and until the rise of Christianity sacrilege was scarcely an issue. Nor is there any reason to suspect a sudden disappearance of career opportunities at least for 'Athenian' philosophers. For besides nature-knowledge, philosophers have always been ready to offer political advice and worldly wisdom, the teaching of which was, and is, always in demand. For an explanation of why decline in natural philosophy occurred when it did, we should instead consider its contents. The Sceptics marked a kind of end point. Their final conclusion (incontrovertible at least in their own eyes) – that all judgement should be suspended – seems to bring the entire adventure of Greek thought to an end. But even for those who did not go along with this, and who continued to accept the Athenian mode of nature-knowledge as valid, it cannot have been particularly obvious in what direction it could possibly be developed further – the supply of suitable first principles seemed to have been exhausted.

So much for 'Athens'; for 'Alexandria' things were different. Until well into the seventeenth century, the flourishing of mathematical nature-knowledge would depend on whether or not rulers were prepared to support it. Some princes were well disposed to keeping mathematicians at court and providing them with an income in return for their services. But it was never certain, and the death of a generous prince could mean downfall for the court mathematician. This may well have been what happened in Alexandria. The Golden Age of Alexandrian nature-knowledge began with King Ptolemy I. He was determined to make his new city of Alexandria the cultural centre of the world that had been opened up by his late commander Alexander the Great and by doing so make Greek rule over Egypt acceptable to his subjects. He founded the famous library, packed with manuscript scrolls collected from everywhere, and the House of the Muses ('Museum'). He set out to attract talent from many cultural fields that, certainly in the case of mathematics, would otherwise have remained untapped. We do not know why he and his successors found mathematical nature-knowledge worth so much money and effort, but it seems fairly clear that, besides adding to the glory of a newly founded dynasty, expert astrological advice was also a welcome ingredient. Neither do we know which of Ptolemy I's successors decided to stop supporting mathematical nature-knowledge. We can do no more than guess that support ceased around 150 BC and suspect that three centuries later Egypt's Roman governor temporarily resumed the practice of his Hellenist predecessors.

Or could Ptolemy (the astronomer) have quite exceptionally been able to live off his many-sided mathematical pursuits without receiving any material or other support from anyone?

For want of hard facts, we can only speculate about our first question. We have more to go on with the second, about the nature of the decline. If nature-knowledge was still being pursued but hardly any really original work was being done, what then was actually being achieved?

The main concern was preservation and in first instance quite literally so. Bear in mind that texts were vulnerable, written on perishable papyrus scrolls, all copied by hand. Many texts had already been lost during classical antiquity itself, and of many ancient writers we now only have fragments. What does remain we owe to the copying, editing and diffusion of texts – seemingly humble tasks but indispensable. As regards content, preservation took many different forms, some more creative than others. In natural philosophy the once so fruitful juxtaposition of exciting new truths gave way to the cease-less repetition, full of hollow rhetoric, of standardised viewpoints from one school or another. There is also the rearrangement and simplification of established ideas for teaching purposes, as well as attempts to reconcile the four main schools by blending a selection of their doctrines. At best, some well-considered variants were produced such as Plotinus' spiritualising of Plato's doctrine, or Proclus' Platonic reflections on the fundaments of Euclid's geometry. We also find expositions of, elucidations of and at times even critical commen-taries on the doctrines of one Athenian school or another.

Meanwhile, the 'nature' portion of philosophy as a whole dwindles steadily. The natural philosophy of the Stoics was virtually lost (and has only been reconstructed with great difficulty from some early fragments), all while their political and ethical doctrines con-tinued to be cultivated. There is also a kind of changing of the guard. The schools where nature featured most, those of Aristotle and the atomists, faded increasingly into the background. Under the late Roman Republic and the early Empire, the Stoa came to the fore while in the late Empire the neo-Platonism of Plotinus predominated, from which some elements were borrowed by early bishops to adorn their Christian message with scraps of classical learning.

As well as repetitions and summaries, elucidations and commentaries, mixtures and variations, there were also translations. The division of the Roman Empire into a western (Rome) and an eastern (Constantinople) empire was reflected in two streams of translation. One was from Greek to Latin, which begins in the first century BC and continues until the sixth century AD. It is accompanied by at times radical restructuring and simplification; what you get is not so much a literal translation as a Latin paraphrase. The other, from Greek into Syriac and/or Persian, was the work of supposedly heretical Christians who were driven out of Constantinople and fled to Persia between the fourth and sixth centuries. Here the translations were much more literal and the original knowledge structure of the texts, whether Athenian or Alexandrian, was maintained. As we shall see, it is this latter stream of translations, centuries after the decline of ancient civilisation, that was the starting point of what would become the Golden Age of nature-knowledge in Islamic civilisation.

So far we have considered Greek nature-knowledge in terms of its own character and achievements. But in this book we are ultimately concerned with the origins of modern science, of which the Greeks are so often seen as the immediate predecessors. That is true, at least to the extent that modern science was grafted in the end on to Greek nature-knowledge and not on to the Chinese. The question we must now ask is why that is so. We shall therefore consider the Chinese approach to natural phenomena and compare it with the two Greek ones.

THE WAY AND ITS SYNTHESIS

We can divide Greek and Chinese nature-knowledge into roughly similar periods. Both have an overture in which, however vaguely, numerous themes are sounded that later become prominent. For the Greeks it is the period of the pre-Socratics (c. 585 until c. 400 BC).

In ancient China, thinking about how the natural world is constituted blossoms in the era of the Warring States, which begins in 480 BC and ends in 221 BC with China's unification under the first emperor.

In both cases, the large number of early themes are subsequently sifted and systematised.

For the Greeks, this takes shape as the founding, in Athens, of four schools of philosophy, each of which draws on a particular pre-Socratic idea. Independently of this, in and around Alexandria a specifically mathematical approach is developed which draws on the pre-Socratic idea of mathematical proof. This whole period of upswing culminates in a Golden Age in which nature-knowledge is pursued intensively and creatively. It lasts for about one and a half centuries, ending halfway through the second century BC. Decline is as sudden as it is steep. The preservation and dissemination of previously acquired knowledge now become the main activity until, many centuries later, ancient civilisation collapses for ever.

Among the Chinese only a limited number of the many leading early ideas survive to be absorbed into a synthesis under the first stable imperial dynasty, the Han. It was the expression of an emergent consensus among scholars at the time. In spite of later variations and expansion, the core and many details of the synthesis were maintained to the end of the empire at the beginning of the twentieth century.

From the start, the central question in Chinese thought was how to bring about and maintain a stable social order. It would have to be in harmony with human nature and this in its turn would reflect the harmonious order of the cosmos. If a coherent, underlying relationship connecting the infinite multiplicity of natural phenomena can be detected at all, it resides in their interdependence. This Chinese mode of thinking has been called 'correlative':

> For the ancient Chinese, things were connected rather than
> caused ... The universe is a vast organism, with now one
> component, now another, taking the lead at any one time, with all

the parts co-operating in a mutual service ... In such a system as this, causality is not like a chain of events ... [rather] succession was subordinated to ... interdependence.

The world is an infinitely fine fabric; each tiny thread is interwoven with all others and the way to grasp it intellectually is to think in terms of correlations and interdependence. Four key concepts of what ultimately became *the* Chinese world-view bring this correlative mode of thought to creative expression: *tao* (the Way), *chi* (matter/energy), *wu-hsing* (the five phases) and *yin–yang*.

Here 'tao' does not refer specifically to the Taoist religion that developed over centuries from one of the many traditions rooted in the hallowed ancient texts from which the Chinese world-view first arose. 'The Way' in its wider sense was the quest of all of the old traditions. It was Confucius (sixth/fifth century BC) who first used the ordinary word for 'way' or 'path' in the sense of the appropriate way of life for individuals and society that is in harmony with the fabric of nature. The sages of old had followed the Way spontaneously, but it now had to be rediscovered. Confucius sought the Way primarily in the observance of proper ritual. The natural order as such did not greatly concern him and the contributions to natural research of his followers and devotees always remained very limited. The two authors whose writings provided the basis of the later Taoist religion distinguished between a Way about which one can speak, the ever-changing Way of natural processes, and the unchanging Way that cannot be spoken of because (as in all forms of mysticism) no words are sufficiently adequate. Many other 'Ways' were advocated, even though there was broad agreement that only one Way could be the correct one. Any Chinese thinker in the period of the Warring States who presented himself at court with advice for the local prince had to be sure to offer a unique narrative that would expound in a persuasive manner the only true Way. But after unification, in the first century of the Han dynasty, the Way was assembled and put under a single label such that it henceforth coincided with the cosmic order in which the new centralised empire was also rooted.

How can one recognise that cosmic order? Understanding is achieved not only intellectually, but also by intuition, contemplation and the powers of visual imagination:

> Study was one of several kinds of self-cultivation. It provided understanding and useful knowledge of the world (which was one aspect of the Way). The deeper aspect of reality (the nameless Way) is so subtle that one can penetrate to it only through noncognitive means.
>
> The Book of the King of Huai-nan puts it cogently: 'What the feet tread does not take up much space; but one depends on what one does not tread in order to walk at all. What the intellect knows is limited; one must depend on what it does not know in order to achieve illumination.'

The other core concepts also go through a process of development of their own before they are moulded together by the Han synthesis into the tools needed to grasp the world fabric.

In so far as 'chi' can be translated into a modern concept at all, it is best rendered by means of an artificial compound term, 'matter/energy'. Originally it represented a disparate series of phenomena: air, breath, smoke, mist, clouds. What they all have in common is that though observable they are not tangible. From there, 'chi' grew to stand for bodily vitality and the climatic and cosmic forces that help to determine health.

Characteristic of 'yin' and 'yang' is that, although they are polar opposites, they also complement each other; even opposites cannot for a moment do without each other but have, as it were, an inbuilt interconnectedness. In principle, every spatial configuration and every process that take place in real time are open to a yin–yang interpretation, whether it involves action and reaction, male and female, or growth and withering.

Wherever duality appears, yin–yang provides an obvious framework. But if more subdivisions are required, wu-hsing, the 'five phases', come in handy. They are represented in material terms:

Water, Fire, Wood, Metal and Earth, which indicate not only the physical substances themselves but also the more abstract processes that regulate the course of things. Thus, Earth represents vegetative processes, Metal those solid figures that can be transformed into another state by melting or by evaporation. The explanatory power of the five phases lies primarily in a distinction being made between various cycles, in particular the Production cycle and the Conquest cycle. The Production cycle 'Wood – Fire – Earth – Metal – Water' indicates the order in which one process flows from the other. And the Conquest cycle 'Wood – Metal – Fire – Water – Earth' marks the order in which one process gains the upper hand over another. Whether such processes are thought of as concrete or abstract, they are not rectilinear but indeed cyclical: after the wooden spade has subdued the earth, the metal axe has carved the wooden object, the fire has melted the metal object, and water has doused the fire, the victorious water in its turn is dammed up and channelled by earth.

These then are the core concepts of Chinese thought, which in the first century BC were moulded into a synthesis. We shall first look at the political developments which made it possible and then at the synthesis itself. Finally, we shall take a look at one highly promising 'Way' which, like the others, dates from the time of the Warring States but was ultimately excluded from the imperial synthesis.

Unlike Greece which, in its innovative period, consisted of a loose collection of city states, China was a compound of medium-sized states constantly at war with each other for hegemony. The period which later became known as the era of the Warring States was a time of outstanding creativity in which the concepts that we have been looking at took shape in a variety of text lineages, later known as the 'Hundred Schools of Thought'. The pioneers and followers of such a text lineage competed with each other to spread their own Way, which meant persuading one of the many princes to adopt it as his guide for political action.

The unification of China under a single emperor was decisive in how the Chinese view of the world took shape. The new absolute ruler who proudly named himself 'the First Emperor' made a determined start by demanding total obedience throughout the empire for no better reason than that he had emerged victorious from a centuries-long conflict. The only expertise that his handpicked royal advisers needed concerned the techniques of the exercise of naked power. Subtler, less crude ways of thought did not suit him, and what survived from the various text lineages of the Warring States era is only what escaped the book burnings or other forms of destruction of that time. In particular, they were to prove fatal for the texts in the tradition of the pacifist Mo Ti (about whom more later).

But no empire can be held together solely by force or the threat of it, and nobody could be more aware of this than those who rose up against the successors of the First Emperor and in 207 BC established the new Han dynasty. If it was to last longer than the brief fourteen years of its ferocious predecessors, it would have to introduce and spread a central conception capable of persuading subjects to accept its authority as legitimate and to obey it freely, not solely under compulsion. It was at this time that what had survived of the old text lineages proved their usefulness. Under the Han, a consensus emerged which on the whole has been characteristic of Chinese thinking about the world ever since. Central to it is the idea of harmony. There is a heavenly harmony which, when things go well, is reflected in the harmonious state of affairs in the empire. Whether things will in fact go well depends on the emperor, whose primary task it is to establish harmony and maintain it. He rules by the Mandate of Heaven and when he loses or forfeits that mandate a change of dynasty takes place. Natural phenomena such as earthquakes or floods may be signs or portents of such a change.

But natural phenomena are more than just signs and portents for humanity; how they inter-relate can also be studied. That inter-relationship, too, is characterised by harmony, and the basic tools for tracing it are to be found in the interplay of the various key concepts that we have just looked at.

The Han synthesis has been called a 'philosophy of organic materialism'. It is a world-view in which material processes, which at the same time have a spiritual component, are taken up in a single organic fabric. Very briefly, it boils down to this:

> The Chinese cosmos is a constant flux of transformation, always regenerating itself as its constituents spontaneously change. Ch'i is matter, transformative matter, always matter of a particular kind, matter that incorporates vitality.
>
> Yin–yang and the five phases had, by the end of the first century BC, a consistent, dynamic character as part of the ch'i complex. Anything composed of or energized by ch'i is yin or yang not absolutely but with reference to some aspect of a pair to which it belonged and in relation to the other members ... Yin–yang provided a flexible language well suited to discussing the balance of opposites. This was a balance not of quantity but of the dynamic quality of each in interacting domains – for instance, something could be yang in its activity and yin in its receptivity. When the focus was not on a binary opposition, however, but on more complex sequences of growth and decay or conquest and subjugation within a larger process, the various sequences of the five phases came readily into play.

What does all this mean in practice? How did scholars actually approach a coherent series of natural events? One would rarely, if ever, encounter all the key concepts in a single specific case. But the approach was marked by three main characteristics: (1) a strong orientation toward practice; (2) a tendency to make classifications in accordance with certain pre-existing schemata; (3) regular interaction between, on the one hand, experiential knowledge and, on the other, the prevailing view of the world as it had evolved during previous centuries into broad consensus.

One example is the discovery in the first century CE of the connection between the tides and the phases of the Moon. The spectacular alternation between ebb and flow in the mouth of the Yangtze river supplied the observational data. But what could have inspired an observer at such an early stage in the history of human thought to correlate tidal movements with a seemingly unrelated phenomenon, the waxing and waning of the Moon? The receptiveness of a Chinese scholar to this kind of connection was furthered by an underlying world-view in which everything is part of an organic fabric and events follow a cyclical pattern.

Another example is the invention of the magnetic compass in the eleventh century by Shen Kua. It had its roots in the doctrine of the suitable building plot (*feng shui*, which has lately become fashionable in the western world) and it was closely tied up with careful observation, of magnetic declination in this case (the difference between 'true' north and 'magnetic' north).

The same Shen Kua knew of a spring whose water, if heated, turned into bitter alum. Further heating produced copper and, if you heated the alum for long enough in an iron pan, the pan itself turned into copper. To the modern chemist this is a familiar substitution reaction, whereas for Shen Kua it fitted seamlessly into the 'Conquest cycle' of the Five Phases: Metal subdues Water. Again, the availability of such a conceptual framework made the researcher receptive to this kind of subtle observation.

Or take sympathetic resonance. If you pluck a string or blow a pipe it may happen that another string or pipe nearby will with seeming spontaneity produce the same sound, a phenomenon used for the fine tuning of bells. Once again practical application was not all there was to it; the phenomenon was also explained, in that *chi* was invoked as a kind of cosmic wind that blows through the pipe or helps to spread the sound of the plucked string. Time and again the notion of mutual harmony in the fabric of nature supported the awareness of such phenomena.

All in all, the broad world-view of the Chinese furthered its strong orientation towards careful observation. Not for nothing was 'the myriad things' a standing expression. But how to avoid being overwhelmed by them? One way out was to impose order on them through classification. For duality there was yin and yang; for groups of five there were the five phases. But there were also other possibilities. For musical instruments, for instance, all kinds of subtly different timbres could be distinguished (timbre is the difference you hear between one particular sound and other sounds of the same pitch and volume). Classification was based on the material out of which an instrument was made, but also according to the eight wind directions (north, north-east and so on). The underlying connection between the timbres was to be thought of once again in terms of *chi*.

Throughout the pre-modern history of China the myriad things were dealt with in the broad fashion just described. The single exception to this is a few surviving, rather fragmentary texts from the tradition of Mo Ti known as the 'Canon and Expositions'. Their approach does not reflect this kind of correlative thinking based on the idea of the cosmos as an organically coherent fabric. The thinking style is more rigorous, the structure is along lines of 'if this, then that' rather than the 'both–and' assumptions of all-inclusive interdependence. In the 'Canon and Expositions' the emphasis is on subjects which would receive but little attention in the Chinese tradition: motion and force; light and shade. For instance, a distinction is being made between cases when the image in a concave mirror appears reduced and right side up and when it appears enlarged and inverted. Such subjects do not reappear once the Han synthesis and its organic materialism have been established; the Mohist text lineage drops out for ever.

GREEK AND CHINESE NATURE-KNOWLEDGE COMPARED

The seventeenth-century sociologist of science Francis Bacon had a splendid metaphor for the acquisition of nature-knowledge. He distinguished three methods: those of the ant, the spider and the bee. You should not, like the ant, patiently keep bringing in one load of material after another – that is the way of the empiricist and it will not get you very far. Neither should you, like the spider, weave a cobweb out of your own body – that is the intellectualist approach, which is no better. On the contrary, like the bee, you should collect nectar from the flowers and turn it into honey in the hive. In other words, merely collecting facts is not yet true science; sheer conceptual schemata are not yet true science; real science consists in bringing the two together in productive interaction.

Bacon's image of the spider was aimed in particular at Greek nature-knowledge. He hoped to persuade his contemporaries to cease following the example of the Greeks who were so little inclined to give natural phenomena a chance to speak for themselves before subjecting them to some philosophical schema or mathematical operation. The comparison to the ant was aimed at others of his contemporaries whom we shall meet later. But for the moment we shall borrow Bacon's imagery to mark the contrast between the Greek and Chinese modes of nature-knowledge.

Not that the contrast is in any way absolute. Certainly, Bacon's characterisation of the Greek approach is very perceptive. Philosophers such as Parmenides and Plato, but also the Alexandrians, habitually rode roughshod over natural phenomena and their messy details. But there are also cases (especially in Aristotle's observations on animals) where there was enough room to explore specific phenomena to some extent in their own right. Conversely, as we have seen, the Chinese investigation of nature was certainly not limited to an unthinking accumulation of facts and details. Against a broad conception of the world as a finely spun organic fabric, bold conclusions were certainly drawn, which could also have useful and practical applications. Nevertheless, taken as a whole, Greek nature-knowledge can fairly and illuminatingly be described as primarily intellectualist, and Chinese as primarily empiricist.

What this highlights in particular is how far removed both these ancient modes of nature-knowledge were from modern science. Neither the Greek spiders nor the Chinese ants can be said to have been on the way to producing that special kind of honey which constitutes modern science. By this I mean the construction, *followed by ceaseless renovation,* of quantitatively oriented models of empirical reality, which have an inbuilt potential for systematic feedback and adjustment such as is made possible, most notably, by experiment. This is only by way of a provisional definition; I shall of course be returning to this cardinal point many times.

It follows that we should judge these pre-modern modes of nature-knowledge first and foremost on their own merits, not according to whether or to what extent they foreshadow what we know would ultimately emerge from only one of them, namely modern science. Looked at without the benefit of hindsight, both ancient Greece and ancient China produced inventive thinkers who boldly attempted to come to grips with a world that stretched out unexplored in front of them. Each in their own way sliced up that world into manageable pieces and identified underlying patterns. Widely differing modes of nature-knowledge resulted according to how they built up these patterns. Strict, speculative systems or mathematical derivation came from the one, concentration on phenomena against the background of a broad world-view from the other. And yet in principle both were of equal value. The fact that modern science was grafted in the end on to the Greek variant and not the Chinese ought not to matter for the question of what each was worth in its own right.

But of course it *does* matter for the question of how it came about. Is it wholly by chance that there was no Chinese Galileo or a Chinese Newton? The question has often been asked and there is no shortage of answers, which unfortunately often degenerate into a kind of parlour game. Everyone has their favourite explanation, whether it is asserting that the Chinese are incapable of coherently structured thought or that Chinese bureaucrats were determined to smother any desire for knowledge. Irritation at this kind of ill-informed frivolity has led many a China expert to declare the whole issue to be insoluble or even pointless. But that is taking things too far. True, it makes no sense to select one or other part of a civilisation and then, in view of its absence elsewhere, to conclude that modern science could never have emerged there. But it does make sense to investigate certain basic conditions within different civilisations and compare these with each other. For us this involves above all the question of how radical innovation comes about – innovation in general and innovation in nature-knowledge in particular.

Civilisations may clash, but they may also cross-fertilise. Take the two great waves of innovation in Greek nature-knowledge, the pre-Socratic overture and, centuries later, the eruption of mathematical nature-knowledge that rapidly became an important constituent of the Golden Age. The first pre-Socratics lived in precisely those places on what was then the east coast of Greece where trade was conducted on a large scale with peoples from further east (think of Babylonian mathematics and how the Greeks picked it up and then raised it from ingenious recipes to mathematical proof). Centuries later the conquests of Alexander the Great overturned the established order in the eastern Mediterranean and brought about all kinds of contacts and interactions which previously would have been far from obvious (think of the founder of the Stoa in Athens, an immigrant from Cyprus). Furthermore, in Alexandria a completely new cultural centre arose (think of Euclid and all the other great mathematicians).

An influx of foreigners and the introduction of ideas and practices of a different kind greatly increase the chance of renewing ideas and customs that have grown stale in the daily grind. It is one of the great sources of novelty and creativity in history. Admittedly, cultural exchange does not automatically stimulate innovation; there are more than enough instances where civilisations just clashed or the interaction came to nothing because their mutual differences were too great. But in the history of pre-modern nature-knowledge, innovation brought about by interaction has been the rule. Or, more precisely, those interchanges have as a rule led to innovation when they took the form of what from now on I shall call 'cultural transplantation'. By that I mean a particular kind of event that stimulates innovation or even transformation of the cultural heritage, where a certain coherent body of ideas and concepts and practices that has been developed in one culture is transplanted to another and falls on fertile soil. In the course of history *this kind of event has happened more than once with Greek nature-knowledge, but never with the Chinese.*

That is no accident. In the past, it has always been military events that provided the unintended and unasked-for impulse for the transplantation of nature-knowledge. The first occasion, which saw Greek nature-knowledge being parachuted into Baghdad, was a consequence of the conquests by the early caliphs and their first civil war in about 760. The second, in twelfth-century Toledo, resulted from the Spanish *Reconquista*; the third, in Italy, from the conquest of Byzantium by the Ottoman Turks in 1453. But in the case of China there was never this type of salutary confrontation between its own body of thought and that of a truly different civilisation. Because the Chinese empire always remained a self-reliant, independent entity, Chinese nature-knowledge never found itself in the Greek position of losing its home and finding itself in need of shelter elsewhere. As a rule, barbarians were successfully kept outside the gates and, when on occasion they did break in, as did the Mongol and the Manchu armies, the new courts quickly adopted the culture of their freshly acquired Chinese subjects. In short, there is an unbroken continuity in the history of Chinese nature-knowledge which was as admirable as, in the long term, it was unfruitful. At least in this domain, thought went round in circles and, even if those circles were quite wide, a circle that imprisons thought remains a prison.

This applies in the first place to the world-view of organic materialism. Under the Han it emerged as the great synthesis of competing views from the era of the Warring States. True, that world-view expanded and became more detailed and ever subtler over time, yet at no time did the occasion arise to reconsider or question its fundamentals. Together with the 'correlative' mode of thinking, of which organic materialism was the expression, it offered a rich and to some degree plausible view of the world. The more one gets used to that way of thinking, the more attractive it becomes, up to the point where one might perhaps wish that the natural world really is as the possessors of the Way fancied it. It has even been argued that the Chinese world-view might, under favourable circumstances, have developed into a variant of modern natural science, though surely a more organically coloured variant. More sensible than this kind of loose speculation is the observation that organic materialism remained firmly locked up within itself and therefore never had a chance to show what it might produce in a process of cultural transplantation.

The best evidence of the kind of latent development potential that remains unused for lack of transplantation is the fate of the text lineage that goes back to Mo Ti. These texts dealt with precisely the enquiries and doctrines that, in retrospect, come closest to what would have been needed for the rise of natural science in the modern sense. They are comparable in that regard to the mathematical science of Alexandria. Invisible to contemporaries and only detectable in retrospect, each carries inside itself a certain development potential. What they also have in common is a marginal position in the culture at large. Alexandrian nature-knowledge was a highly specialised court product with hardly a life of its own outside the court; of the entire Mohist tradition only a single, fragmentary text lineage remained by the middle of the Han dynasty in the first century CE. But the great difference between the two lies in the circumstance that after the Han the Mohist tradition was swept off the stage completely *and for ever* by the great, imperially approved synthesis of the Way. It was not revived elsewhere nor could it be. The Alexandrian tradition, by contrast, which had also lived on the margins and finally fell into decline with ancient civilisation as a whole, nevertheless received two chances of revival, in Baghdad and once again in Renaissance Italy. We shall soon see how productively those opportunities were actually used.

DEVELOPMENT POTENTIAL AS CORE EXPLANATION

'Cultural transplantation', 'transformation', 'latent development potential': these are the concepts that I shall be using to explain how, under comparable circumstances, one body of ideas and practices may result in something more or less radically new, while another does not.

The idea of 'latent development potential' goes back to Aristotle. He saw it as an appropriate way of circumventing Parmenides' paradox: change is the fulfilment of a potential already present in whatever it is that changes. He put the idea to work for coming to grips with the natural world (with great success at first, but ultimately in vain). We now put it to work for coming to grips with the course of history. To be sure, one big difference is that Aristotle's concern was with the realisation of certain ends: the oak tree which starts out as an acorn is already a hidden presence in the end-state of the acorn. But in history things develop differently. Ends are laid down by humans, some of which are achieved, some not, but the course of history as such is not moving towards some end already pre-determined by Someone or something – the outcome is open. Outcomes of historical processes, including the rise of modern science, are in every case the consequences of a shifting combination of coincidence on the one hand and, on the other, chains of events that hang together with a certain inherent logic of their own which can be investigated with a view to what may have caused them.

What I mean by 'latent development potential' and what it clarifies are best illustrated by a concrete example. A common method of measuring time in the Old World made use of a steady outflow of water. However, the weight of water when the vessel is full makes the water flow out quickly, but as the amount of water diminishes the flow slows down. At the end of the eleventh century a certain Su Sung constructed a complicated water clock. By hanging thirty-six buckets from an enormous water wheel and making them pour out regularly he managed, to a large extent, to equalise the outflow of water.

Modern reconstructions show that Su Sung's water clock would have told the time a great deal more accurately than the mechanical

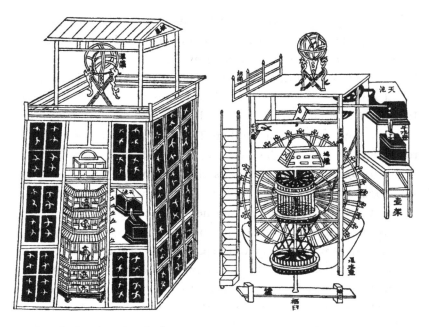

Su Sung's water clock
The whole apparatus was about ten metres high. The cut-away view on the right shows the main features of the mechanism. The illustration on the left shows its purpose: to drive the 'armillary sphere' at the top (the rings represent the meridian, the ecliptic and the celestial equator) and the heavenly globe below (half sunk into the wooden casing). On each level of its five floors, a figurine indicates time.

clock that appeared some two hundred years later in Europe and that operated on a quite different principle, not outflow but oscillation.

The superiority of Su Sung's clock can hardly be doubted. Early mechanical clocks still needed a sundial to check the time indicated: they could be wrong by up to a quarter of an hour a day, whereas Su Sung's water clock varied by a minute at most. And yet its superiority was highly precarious. In the long run, this ingenious water clock begins to jam and ends up entirely stuck. Because of temperature variations, rust and the accumulation of dirt, the regularity of the outflow gradually diminishes. Furthermore, the design does not allow for modification, nor could the giant water wheel be reduced in size. And so it remained an isolated incident; throughout the history of Chinese civilisation fewer than half a dozen such clocks were built. The American historian David Landes strikingly called Su Sung's clock 'a magnificent dead end': his water clock exploits to the full a principle that does not allow of any further development. It stood at the end of one technical development, the mechanical clock at the beginning of another. The mechanical clock could be reproduced in large numbers without great difficulty, it could easily be repaired, and working variants could be made, which were far smaller and even portable. Above all, there was the potential for drastic transformation, the replacement of the foliot by a genuinely uniformly swinging regulator, the pendulum. This only happened centuries later when, as one of the early fruits of the rise of modern science, the properties of the pendulum were discovered. From then on the deviation each day was reduced to no more than a few seconds.

The mechanical clock, in short, rested on a principle that bore within itself a potential for further development in ways that nobody knew or could have known about. Landes has used this in respect of the two distinct principles of measuring time; I have come to see it as the best possible path towards explaining the rise of modern science.

An explanation along such lines takes the form of a threefold 'if'. A specific body of nature-knowledge emerges as the product of a period of impassioned discovery lasting several centuries. There are now three successive possibilities:

(a) The corpus of knowledge either does or does not carry within itself a substantial potential for development. If it does not (as may well have been the case with Chinese nature-knowledge), we are facing a magnificent dead end; whereas if it does, the following possibility arises:

(b) The corpus may or may not be subjected to processes of cultural transplantation. If not (as was certainly the case with the Chinese body of nature-knowledge) it will remain locked up within itself,

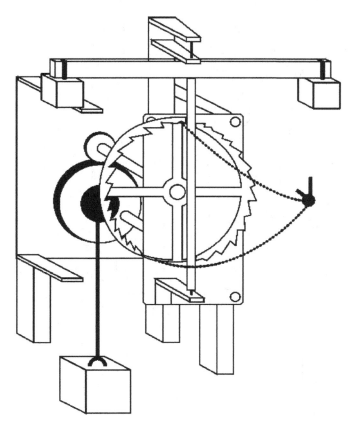

The mechanical clock
By means of a cog mechanism controlled by a metal rod (foliot)
with small movable weights, the descent of a suspended weight is
held back, converted into a to-and-fro movement, and made uniform
as much as possible. Uniformity is actually quite deficient due
to the decreasing regularity of the to-and-fro movement as the weight
descends.

forever true to its origins and unchanging at its core. But if it does
undergo such a process, a final possibility comes in sight:

(c) The corpus may or may not be transformed into something more
or less radically different.

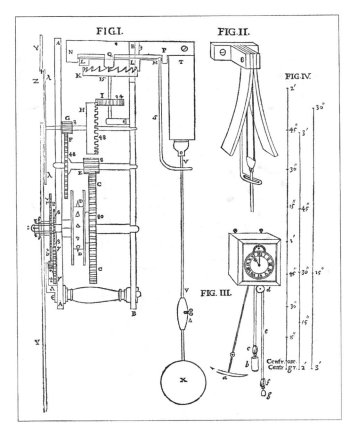

Huygens' pendulum clock
Below right (Fig. III) the clock; left (Fig. I) the mechanism shown from
the side; above right (Fig. II) how the pendulum is suspended and
attached to the wheel mechanism.

With (a) the crucial question is at precisely which point the phase of enthusiast discovery comes to an end, as happened with Greek nature-knowledge in the second century BC. When decline sets in, does the corpus still possess fertile possibilities for development? If so, it is best to regard it as a gift from history. Naturally, praise or blame in respect of those who happen to receive the gift is not at issue here. As historians, we can do no more than record the event, in full awareness that we can do so only in retrospect.

With regard to (b), whether or not cultural transplantation does occur will depend on a range of factors which have nothing to do with the specific nature of the body of knowledge in question. One of them is geographic location, in particular the presence or absence of centres of inter-ethnic exchange. Decisive, however, is the absence or presence of large-scale military conquests, which may shake up an entire civilisation or even give rise to a new one.

The fate of Greek nature-knowledge in Byzantium provides some negative evidence for the decisive impact of cultural transplantation. Towards the end of the Eastern Roman Empire, but still before it became the independent state of Byzantium in the eighth century, texts containing Greek nature-knowledge were gathered and piled up in the palaces and monasteries of Constantinople. They were, naturally, written in the original Greek and were immediately accessible to any Byzantine scholar who wished to consult them without having to master any necessarily complex process of translation. And consulted they were, and also copied carefully in beautiful handwriting on to parchment, while some were even partially adapted and expanded a little. Even so, we are bound to observe that, for almost a thousand years until the fall of Constantinople in 1453, virtually nothing creative was done with Greek nature-knowledge. Things had just been made too easy for the Byzantine scholars to whom it had all been handed on a plate. What was missing was precisely the challenge involved in a process of cultural transplantation.

Finally to point (c), which raises the possibility of a more or less drastic, perhaps even revolutionary, transformation of the transplanted nature-knowledge. In the following chapter, we shall see that this possibility has manifested itself two or three times in the past and that on one occasion the potential became reality. The way in which this happened resolves the enigma – in so far as it can be resolved at all – of how around 1600 a small band of European scholars, who had grown up in the Greek tradition, were able, as it were, to re-create the world so thoroughly that four centuries later we have still not seen the last of it.

2 Islamic civilisation and medieval and Renaissance Europe

I shall chart the successive transplantations of Greek nature-knowledge by way of a metaphor. Or, rather, let us extend in a literal sense our image of transplantation. Imagine that in Greece a flowering shrub, an oleander for example, has been sown and carefully cultivated. Centuries later the nursery goes bankrupt. The administrators (in this case the rulers and scholars of Byzantium) are not particularly concerned whether the plant has withered or not but luckily it can survive for centuries without water. However, if it is to recover fully, cuttings will have to be taken and it will certainly need repotting. Also, customers will need well-tilled land in which the cuttings can take root, as well as compost to help it to blossom.

On three occasions, a customer did come to the door and, in each case, his land had been thoroughly ploughed and prepared. However, the ploughing had been done by warfare.

Warfare is obviously horrendous for the many who throughout history have been its victims. Daily routine is lost and nobody's life is safe. But, by turning everything on its head, war also creates space for change. Such change can be large-scale and at times even creative. We have seen how in the Greek world the conquests of Alexander the Great led to the founding of a cultural centre where certain pre-Socratic ideas were systematically developed into a mathematical mode of nature-knowledge.

The creative potential of warfare is evident on each occasion that cuttings are taken from Greek nature-knowledge and planted elsewhere. The first customer who came to the nursery was a caliph, some way down the family line of the prophet Mohammed. His name was al-Mansur and he came to power around the Islamic year 140 (or *c.* 760 AD, according to the Christian calendar which I shall keep using for the sake of convenience). His accession had not been straightforward. His family had engineered a coup and had won the civil war that ensued. He justified the coup on the grounds that he was a descendant of the prophet's uncle Abbas, and the dynasty that he founded is known accordingly as the Abbasid Caliphate. Determined to make a fresh start, he began by founding a new capital city of Baghdad on the model of Alexandria. The similarity went much further than its grid street plan. Al-Mansur gave orders that ancient Greek manuscripts should be collected from throughout the Islamic world and brought to Baghdad to be translated into Arabic. Most of them had to be translated from Syriac or Persian, but emissaries of the Abbasid caliphs also went to Byzantium and brought back original Greek texts.

What was it that inspired al-Mansur and his successors to persist for generations in this undertaking? They certainly knew what they did and did not want. Poets such as Homer and Euripides were ignored; Aristotle and Ptolemy and other big names of Greek nature-knowledge were sought after. Some of the motivation was the same as with the early Ptolemys: solid astrological advice and legitimation of an authority which (apart from Uncle Abbas) was based on nothing better than conquest. But there was more to it than that. Among their new Persian subjects there was a legend that so-called Greek nature-knowledge was originally Persian but that Alexander the Great had stolen it. By restoring Greek nature-knowledge to its rightful owners, as it were, the caliphs hoped to prepare educated Persians for service at court as well as conversion to Islam. One and a half centuries had passed since Mohammed had embarked on his mission in Mecca in a far-from-civilised world. Leading Muslims were well aware that when it came to culture, as conquerors with the sand of the desert still in their hair, they fell short of their numerous Jewish, Christian and Zoroastrian subjects. They regularly came off worst in theological debates. It was therefore the astronomical and astrological texts of Ptolemy, and a treatise by Aristotle filled with guidelines for effective debating, that were the first to be translated into the language of the

Koran. In the two centuries following al-Mansur's initiative, translators increasingly followed their own preferences, and by the beginning of the tenth century most of what had survived of Greek nature-knowledge from antiquity had been translated into Arabic. The manuscripts circulated from Andalusia on the western border of the Islamic world to Afghanistan in the east. They were also not hugely expensive: in the Baghdad market you could pick up a copy of Ptolemy's *Almagest* for the price of a donkey.

The second transplantation began in the Spanish city of Toledo. This time it was a matter of repotting some cuttings. Greek nature-knowledge in Arabic had spread widely throughout the Muslim world including al-Andalus (Andalusia) in south-eastern Spain. Its capital Cordoba was a cultural centre but now, four centuries after it had been captured by the first caliphs, it was under increasing military pressure from the Christian kingdoms in northern Spain. The reconquest of Spain from the Islamic 'Moors', still referred to by the Spanish term *Reconquista*, was finally completed at the end of the fifteenth century. But Toledo (not far from present-day Madrid) fell as early as 1085, giving rise to another large-scale wave of translations. From the Arabic cuttings taken from the original Greek yet another cutting was taken and planted in Latin soil. The instigator of this was an Italian travelling scholar from Cremona called Gerard. He was searching for the text of Ptolemy's *Almagest* but had been unable to find a copy in Europe's few libraries. So in about 1145 he ended up in Toledo where he set about translating the *Almagest*. He remained there for the rest of his life and translated in total about seventy Greek works from Arabic into Latin. Initially, because his Arabic was not yet very good, he and the dozen scholars who joined him in the enterprise needed the help of Islamic and Jewish scholars – an early, heart-warming example of harmonious collaboration between adherents of the three great monotheistic religions. Once again the texts chosen for translation were mainly about mathematics and about natural philosophy, with an emphasis on Aristotle's works and numerous commentaries on them.

What was it that drove these translators? For a better understanding, we must for a moment transport ourselves back to the Europe of that time. Europe had long been at most only a superficially civilised appendage of the Mediterranean Roman world. Then, in the seventh century, the first caliphs effectively closed off the Mediterranean with their wars of conquest which extended from Persia to Spain. Europe was thereby cut off from the civilised world and from then on had to proceed alone. At the start of the ninth century it was briefly united under Charlemagne. But the court of Charlemagne could never compete in versatility, refinement and cosmopolitanism with that of al-Mansur's grandson Harun al-Rashid. For a couple of centuries after Charlemagne's empire disintegrated, while Norsemen terrorised Europe's coasts, it was only in churches and monasteries that traces of civilisation survived. The same applied to nature-knowledge. While it flourished in the Arabic world, in distant Europe a few monks and priests struggled to make sense of the summary texts that had taken the 'western' translation route. These texts were only a poor reflection of the original wealth of Greek nature-knowledge. Those that had taken the 'eastern' route, via Syriac and Persian, had preserved much more of its richness. Their subsequent translation into Arabic had enabled them to spread throughout the entire Islamic world, even as far as Andalusia.

That is what made it possible for Gerard and his colleagues in the mid twelfth century to discover such a treasure trove of accumulated nature-knowledge. They could drink from what must have seemed to them to be the fount of knowledge itself. And it gave them the motivation, over a period of fifty-odd years, to render Aristotle, Ptolemy, Euclid and numerous other scholars as well as the texts of Arab commentators into their own language of scholarship, Latin.

Another 300 years later in 1453 a third client knocked on the door. He appeared at the nursery just as it had changed hands after a hostile takeover. In the greenhouse he found a plant that was extremely old and completely withered. But he took not just a cutting but the entire plant, crossed the Adriatic and planted it in Italian soil.

That is what happened to Greek nature-knowledge after the conquest of Constantinople by Sultan Mehmed II, ruler of the Ottoman Turks. He made Constantinople his capital city and renamed it Istanbul. But it was not he who was interested in the original plant. In his territories the cuttings of the early caliphs were still, or once again, in full flower. Therefore, the Ottoman practitioners of nature-knowledge did not feel any particular need to concern themselves with the wealth of manuscripts which had now suddenly been released from the palaces and monasteries of Byzantium.

A Byzantine priest named Bessarion did feel the need. He had settled in Italy and as a convert to the Roman branch of Catholicism had eventually risen to the position of cardinal. It was primarily thanks to him that those ancient Greek texts were brought to Italy to become the springboard for a third and final transplantation of Greek nature-knowledge. It could now be translated directly without the distortions inadvertently brought about by all those intermediate steps through Syriac, Persian and Arabic. Direct contact could now be made with the minds that had conceived of those texts one and a half millennia earlier. The excitement at actually being able to stand at the fount of all nature-knowledge was palpable, first of all in Renaissance Italy but in the course of the sixteenth century throughout western Europe. The third transplantation had taken place.

In the course of about one and a half centuries all the work of Gerard of Cremona and his colleagues was repeated (at a higher level, to be sure) and also expanded. By about 1600 a scholar in Europe had at his disposal authoritative publications of virtually every surviving text relating to Greek nature-knowledge. The printing press made such direct access even easier for him than for his predecessors. But it did not stop at translation, any more than it had in Islamic civilisation or medieval Europe. The flowering of the plant and its cuttings went beyond just restoring them to their old form in fresh and fertile soil. New branches appeared and from these new branches were to sprout even more splendid blooms.

MATTERS OF TRANSLATION AND FORMS
OF ENRICHMENT

Translation is more than substitution. It is a highly active process, certainly for texts dealing with nature-knowledge. The translator must have more than just a command of the source and target languages. For the source text he must acquire a thorough grasp of the technical detail, of the philosophical jargon, of the mathematical terminology and styles of mathematical proof; otherwise he will not understand what is meant when the author refers to *pneuma* or to 'parallel'. In the target language when it has not been used before as, for instance, the language of the Koran, he even has to create a new vocabulary. But should he think up new words or give existing words a new meaning? And should he exploit the particular qualities of the target language? Simon Stevin, for instance, at the end of the sixteenth century coined Dutch words to replace (and clarify) Greek-based mathematical terms. They are still in use today in modern Dutch as, for instance, the word for mathematics itself which he called '*wiskunde*', expert knowledge ('*kunde*') of what is certain ('*wis*').

To solve this kind of problem, translators with great creative gifts were required. But something else was required in addition. In those times the modern ideal of remaining as faithful as possible to the original text was not so much a translator's concern. Here is how in the ninth century a mathematician of note described his task in his foreword to the *Almagest*:

> The work was translated from the Greek into the Arabic language by Ishaq ibn Hunayn ibn Ishaq [etc.] for Abu s-Saqr Ismail ibn Bulbul and was corrected by Thabit ibn Qurra from Harran. Everything that appears in this book, wherever and in whatever place or margin it may occur, whether it constitute commentary, summary, expansion of the text, explanation, simplification, explication for the sake of clearer understanding, correction, allusion, improvement and revision, derives from the hand of Thabit ibn Qurra al-Harrani.

In short, a translator aims to put the reader in the best position to understand the most up-to-date aspects of the subject matter dealt with in the translated text which includes providing explanations and glosses. This approach brings together two activities which might appear to be entirely separate: the *translation* of a text and the *enrichment* of its content. With Thabit ibn Qurra this process is well under way: indeed he made important contributions to the enrichment of a number of Greek ideas. For instance, he went beyond Archimedes in his study of solid equilibrium to include the weight of the beam itself. Al-Biruni, one of the greatest Islamic practitioners of nature-knowledge, enriched Archimedes' work on fluid equilibrium with a concept that comes very close to the modern concept of specific gravity. In a similar vein, Thabit and al-Biruni and other mathematical astronomers made corrections to the theoretical models in Ptolemy's *Almagest*.

The problem in the last case was that the observational data in the *Almagest* which had originally matched the models quite neatly were no longer so accurate nearly a thousand years later. Ptolemy had assumed that the duration of the solar year is a constant value; similarly so with the angle made by the ecliptic with the celestial equator. Once it became clear that both the duration and the angle had changed, the question arose of what to do about it. You could ignore the old data and use the new and with the help of the models in the *Almagest* draw up new tables for the positions of heavenly bodies at various times in the future, running the risk that at some point they would again become obsolete. Or you could accept the correctness of both the old and the new data. But that implied a recognition that the duration of the solar year, or the angle of the ecliptic, were subject to long-term variations, making it necessary to adjust the models to take account of it. It says much about the depth of the process of enrichment that both in Islamic civilisation and in Europe a number of astronomers opted for the second, much more challenging alternative.

Sometimes a problem which in Greek antiquity had been left more or less unresolved finally found a solution in one or more of the

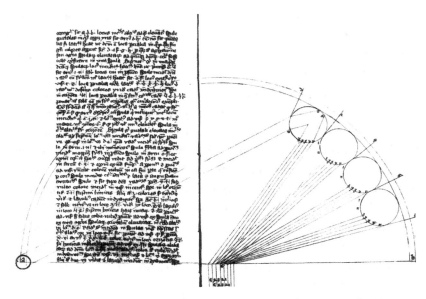

How, according to Dietrich von Freiberg, a rainbow comes about
The Sun is in the bottom left, four raindrops are on the right and in the
lower centre is the observer. Within each raindrop the colour spectrum
arises through refraction, then reflection and then again refraction.

three 'transplantation cultures'. One example is the rainbow. Building
on the work of Ptolemy and Ibn al-Haytham (the best of the Arabic
scholars to continue and extend Ptolemy's work), both Kamal ed-Din
al-Farisi and Dietrich von Freiberg, though living thousands of miles
apart, found out at almost the same time how rainbows occur.

Or take the regularity with which a light ray at successive
angles refracts at the interface between, for instance, air and water.
Ptolemy had worked out a rough approximation. The exact regularity
(known as the sine rule of refraction) was discovered in the tenth
century by Ibn Sahl. His discovery, which he recorded in manuscript,
led a hidden existence. Seven centuries later the same discovery was
made again by a trio of adherents of Alexandrian nature-knowledge,
independently of Ibn Sahl and independently of each other. However,
the manuscript in which Willebrord Snel recorded his discovery

vanished; Thomas Harriot's manuscript, like that of Ibn Sahl, has only
recently been recovered; and only René Descartes actually published
his discovery in print, which included an attempt at a proof.

These are all examples of the enrichment of 'Alexandrian'
nature-knowledge. But there are also instances of enrichment of the
'Athenian' mode of Greek nature-knowledge. One example is
Aristotle's explanation of projectile motion. We have already discussed his concept of 'natural' motion: objects made of earth and/or
water move whenever possible to the centre of the universe, with
which the centre of the Earth therefore coincides. We observe this
whenever an object is dropped vertically. But what happens when we
throw a stone? Aristotle called this 'violent' motion and it caused him
difficulties. Why is it that a stone does not drop straight down as soon
as it leaves the hand that throws it? After all, according to him there is
no movement without a mover. Where is the mover that propels the
stone forward for a while? Aristotle found the solution in the air. The
hand of the thrower passes its motive force on to the nearest layer of
air which in turn passes it down in weakened form to the next layer.
This force keeps the stone in motion until it comes to rest with the
last layer of air on the ground. This elaborate construction shows the
extent to which the knowledge structure of natural philosophy could
prevail over unprejudiced observation. Surely we see and feel that air
actually holds back movement? And indeed some followers of Aristotle agreed. Ibn Sina, known in Europe as Avicenna and one of the
greatest Islamic Aristotelians, suggested an alternative explanation of
a thrown object which was developed further in medieval Europe.
This stated that the throwing hand does not transmit its motive force
to the air, but to the object itself. It thus acquires an inner motive
force, called 'impetus', which sustains the movement of the object for
a while.

This solution fits smoothly into Aristotle's teachings. The way
of thinking and the knowledge structure remain unaffected – he could
have thought of it himself. Similarly with the other examples of
enrichment developed under all three transplantations. At their best,

those additions and corrections are certainly evidence of considerable critical powers and inventiveness. *And yet they left the 'Alexandrian' and 'Athenian' approach and their respective knowledge structures unchanged.* The fundamental assumptions of the Greeks were not questioned and therefore not replaced, even in part, by new ones. Neither was any attempt made to broaden the limited range of natural phenomena to which mathematical nature-knowledge had traditionally been applied. It remained the familiar five: the trajectories of the heavenly bodies, musical intervals, light rays, and solid and fluid equilibrium. Furthermore, with an occasional exception, they continued to be treated with the same far-reaching degree of abstraction as they had been in Alexandria. As for Athenian natural philosophy, rivalry flared up again between the schools which each continued to claim a monopoly of the truth based on self-evident first principles. The Sceptics' objections to all this re-emerged, too, and there were also regular revivals of the various styles of pursuing natural philosophy which had followed the Golden Age of Greek nature-knowledge: summaries, explanations, commentaries, mingling and blending.

Not only did the enrichment of surviving Greek nature-knowledge remain limited in its way of thinking and in its structure, but also in its content. Take the most radical innovation of them all. Greek mathematics had been confined almost exclusively to geometry but in the early Abbasid period a form of algebra was added. In the cosmopolitan atmosphere that characterised Islamic civilisation at that time, Arabs and Persians were receptive to the non-geometrical form of mathematics which they encountered in India. It was there that positional notation was invented in which the value of a digit depends on its place within the number (the 7 in 374 does not represent 7 but 70). This not only made daily calculations much simpler but, with the additional introduction of zero, it also paved the way to a form of algebra. Al-Khwarizmi, the first writer of importance on the subject, distinguished six standard forms of equation (for example, $ax=b$ or $ax^2+bx=c$, though all expressed in words, not symbols). It was a radical innovation with tremendous potential for the future.

But that is being wise after the event. Until far into the Renaissance period, even the most advanced results achieved using algebra hardly went further than what was or could be achieved using the traditional geometric tools of Euclid and Apollonius. And so once again we see a process of enrichment that never really strays beyond the established framework. It applies to Islamic civilisation and equally to medieval and Renaissance Europe. At the beginning of the thirteenth century Leonardo Fibonacci wrote a textbook explaining place-value notation and from then on merchants were able to do their sums much more easily. For nature-knowledge it made no difference. And it was exactly the same with algebra.

Time and again we see that both mathematics and transplanted nature-knowledge could be and indeed were enriched up to the limits of their traditional intellectual framework. *But those limits were never actually transcended.* Nowhere is this so clearly illustrated as in the 'bridging' function which Ptolemy had made such a point of. In direct emulation two attempts were made to link the intensely abstract nature-knowledge of Alexandria more closely to the natural world. In the eleventh century Ibn al-Haytham (Alhazen) took up the challenge. He was thoroughly at home with Ptolemy's work, he was familiar with Greek natural philosophy, and he was a physician. He extended the geometric treatment of light to provide a general explanation of vision, in which he combined a geometric approach to light rays with observations of the eye's anatomy and any scraps of natural philosophy which seemed suitable for the purpose.

Some experiments with a revolving oil-lamp supported his argument. By modern standards, Ibn al-Haytham's synthesis has little value. But that is not what matters for the historian. In his imaginative processing of insights from distinct domains into a single coherent whole, his work belongs to the best that the three cultural transplantations of Greek nature-knowledge produced. Five centuries later, a learned choirmaster of the San Marco cathedral in Venice, Gioseffo Zarlino, undertook a comparable synthesis in the domain of musical harmony. Both attempts at synthesis proved untenable and

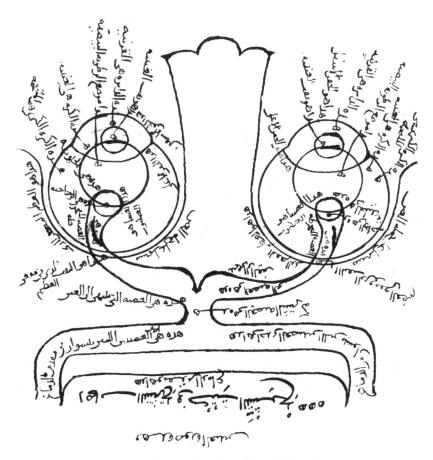

How we see according to Ibn al-Haytham (Alhazen)

were shipwrecked in the early seventeenth century on the rocks of emerging modern science. But that is not the point. The point is that Ptolemy's attempt to link mathematics more directly to the natural world was picked up and worked on creatively in Islamic civilisation and later on again in Renaissance Europe. Ibn al-Haytham and Zarlino deliberately adopted Ptolemy's own solutions as a starting point and continued in the same spirit. Each time, we find ourselves at the outer limits of Greek nature-knowledge but without overstepping them.

At the end of the sixteenth century, anyone looking at how Greek nature-knowledge had been handled in Islamic civilisation and in Europe could see that much had been clarified and expanded but that the core had remained unchanged. In the mathematical tradition it is Archimedes and Ptolemy, and in natural philosophy the Athenian schools, that remain the point of reference for everyone wishing to work in and develop Greek nature-knowledge.

ISLAMIC CIVILISATION: SPECIAL DEVELOPMENTS

So far we have considered what our three cases of cultural transplant-ation had in common in the process of translating and processing the Greek corpus of nature-knowledge. A pattern has emerged of content-enrichment that retains the original knowledge structure. But natur-ally there are also differences.

One obvious difference is in the timing of the transplantations. In Islamic civilisation it began in the eighth century, in medieval Europe in the mid twelfth century and in Renaissance Europe about three centuries later. Consequently, successive phases could benefit at least to some extent from insights gained earlier. Ibn al-Haytham's synthesis of light and vision, for instance, was adopted in medieval Europe and survived unscathed during the Renaissance.

Another difference is that Greek nature-knowledge was not always transmitted or absorbed in equal measure. Relatively speak-ing, the broadest revival was during the Renaissance, while much the narrowest and most one-sided was in the Middle Ages. Islamic civil-isation lay somewhere in between. Although elements of atomism, Stoicism and even some of the Sceptics' pervasive criticisms found an echo in Islamic civilisation, it was Aristotle and Plato who dominated the philosophical debate. Not that a strict distinction was always made between the two. The summaries and commentaries of the greatest of their philosophers, al-Kindi, al-Farabi and Ibn Sina, show a penchant for approaching Aristotelian problems in a Platonic spirit.

In Islamic civilisation knowledge was divided into 'Arabic' knowledge and 'foreign' knowledge. The former comprised knowledge of the Koran, of the tradition of the prophet (*hadith*), and of law and jurisprudence (*sharia*). 'Foreign' knowledge stood for the rest, of which Greek nature-knowledge was a major constituent. Here a further distinction was maintained, not formally but in actual practice, between 'Athenian' and 'Alexandrian'. They were and remained, as they had for the Greeks themselves, two strongly divergent forms of nature-knowledge between which there were scarcely any points of contact or interaction. With one major difference. In Greek antiquity philosophers and mathematicians were invariably different persons, whereas in Islamic civilisation that was not always the case. The natural philosophers, al-Kindi and Ibn Sina, also did mathematical work. But what is striking is that in their writings natural philosophy and mathematical science hardly if at all impinged on each other – a phenomenon that we shall encounter again and again in Europe up to and including the case of Descartes.

Meanwhile, nature-knowledge in the Islamic world was not exclusively concerned with the processing and enriching of 'Athens' and 'Alexandria'. In the margins we find elements that clearly reflect a specifically Islamic origin. The Koran prescribes that every believer should direct his prayers in the direction of Mecca. As Islam expanded beyond the limits of the Arab peninsula, it became increasingly difficult to orient new mosques precisely towards Mecca. Theoretically speaking, it is a complex problem of spherical trigonometry, and Islamic mathematicians took about two centuries to come up with an exact solution (nowadays the *qibla* is done by GPS). The Koran further demands that the faithful should pray five times a day but at times that depend on natural events such as dawn, dusk or length of shadows, which are not always easily determined. A large amount of ingenuity went into solving this problem. The same applies to the rules laid down in the Koran for dividing up inheritances, to which an important part of al-Khwarizmi's pioneering work in algebra was devoted. Furthermore, building a community of the faithful created a need to keep it healthy, which led to the first public hospitals in history. One also sees that many practitioners of

Greek nature-knowledge were at the same time medical doctors. Finally, in Islamic civilisation we encounter the quest for the transmutation of metals or, more popularly, for ways and means to turn lead into gold. From its early origins to its demise in late eighteenth-century Europe, alchemy was the most multi-cultural of those fields in which nature-knowledge was cultivated. The first steps were taken in Alexandria. Chinese scholars then added the idea that the means of bringing about transmutation would also have the power to prolong human life: the 'elixir'. Alchemy is not just a collection of ideas about how metals mature in the earth from coarse (lead) to noble (gold) and how that process can be artificially accelerated. It is also a collection of techniques for bringing about that acceleration by means of combustion, distillation, filtration and similar processes. In Islamic culture all this was shrouded in esoteric imagery in which the purification of substances reflected the purification of the soul.

Three particular features of the transplantation of Greek nature-knowledge to Islamic civilisation have now been discussed. It was the first transplantation; it comprised most, though not all, of the texts; and there were also a number of aspects that were neither Athenian nor Alexandrian but reflected specific characteristics of this specific civilisation.

There is a final important aspect – the decline of nature-knowledge in Islamic civilisation. That there was a decline is not in serious doubt, yet there is very little agreement as to when and how and why it happened. In fact, so many different explanations have been given, many of them frivolous and without much of an empirical foundation, that scholars who are able to read the original Arabic and Persian texts have been inclined to set the whole issue impatiently aside as too speculative for historically responsible answers. Again, their reluctance to face it seems to be taking things a step too far. It is true that single events, if treated as unique, would not as a rule offer sufficient empirical grounds for a convincing explanation. But here as elsewhere historical comparison may come to the rescue. In the present case, a sufficiently comparable event readily presents itself. It is the decline of Greek nature-knowledge in its original form. We have already looked at it as part of a wider pattern: an enthusiastic upsurge, culminating in a Golden Age, followed by sudden and steep decline. I have pointed out two things about that decline: (1) it is followed, even centuries later, by occasional flashes of creative ingenuity (for example, with Ptolemy); (2) to ask why there was a decline is to ask the wrong question.

The broad pattern of rise and fall, upswing and downturn, in both civilisations is tabulated below. The most significant differences are italicised.

Of course, the earlier upswing was different from all three later ones. In the Greek case, a good part of it consisted of original insights

	Greece	Islamic civilisation
UPSWING	creation + (in mathematics) transformation	*translation ⇒ enrichment*
• **Golden Age**	Plato to Hipparchus	al-Kindi to Ibn Sina, al-Biruni and Ibn al-Haytham
DOWNTURN	*c.* 150 BC	*c.* 1050 AD (AH 440)
(1) Why?	'normal'	'normal'
(2) Why at that time?	sceptical crisis? cessation of patronage?	*invasions ⇒ turning inwards*
(3) Aftermath	steep, protracted; generally low level	steep, protracted; followed by *partial, higher-level reversal* sustained by court proliferation and by institutional strongholds (observatory, later *madrasas*)
• **marked by**	codification, commentary, syncretism ...	commentary mostly
• **still outstanding**	Ptolemy, Diophantus; Proclus, Philoponus	Nasir ed-Din al-Tusi, Ibn as-Shatir, Ibn Rushd

and, for the rest, of the transformation of knowledge from elsewhere; in Islamic civilisation the Greek inheritance was subjected to a long process of translation and enrichment. What they had in common was a passion for exploring brand-new territory, a momentum and inspiration radiated by the quest to discover the unknown or rediscover knowledge that had been lost and had to be restored to its former glory. In both cases this early excitement culminates in a Golden Age – a kind of event I have defined as a 'strikingly dense constellation of great creative talents'. In the Islamic world, following a series of Baghdad luminaries such as Thabit ibn Qurra, a high point was reached in the first half of the eleventh century by in particular Ibn Sina (Avicenna), al-Biruni and Ibn al-Haytham (Alhazen). After that, as with the Greeks before them, there is a steep decline. Their work is not picked up, nor developed further or surpassed, or even (who knows) re-examined down to its foundations and transformed into something different. Nothing like that occurs, and with the almost simultaneous death of this famous trio there is no further development for centuries.

Why?

As I have argued above, in the Old World a regular pattern of rise and fall, upswing and downturn, was the obvious natural course of events. What does demand an explanation is the world-historically unique *reversal* of this pattern: the singular permanence that has come to characterise modern science. In itself, then, there was nothing exceptional about the decline of nature-knowledge in the Islamic world; the answer to the question 'Why?' is once again 'But what else would you expect?' Still, the more *specific* question of why it ran out of steam just then at around 1050 does need to be asked even if parts of the answer may remain uncertain. The downturn has often been attributed to the famous sack of Baghdad by the Mongol armies of Genghis Khan's grandson Hulagu in 1258. Also often blamed for the decline is an influential treatise from the end of the eleventh century by al-Ghazali, who on behalf of 'Arabic knowledge' subjected 'foreign knowledge' to a fierce critique. Such attempts at explanation are inadequate. The first, for instance, fails in the timing – the downturn took place long before 1258, and even at the height of the Golden Age Baghdad had already ceased to be a centre of nature-knowledge. Al-Biruni and Ibn Sina served dynasties in or even to the east of Persia, while Ibn al-Haytham worked at the court of the Egyptian Fatimid dynasty. As for the second explanation, however great an impact a certain book may have, no single book on its own can bring to a standstill a centuries-old enterprise to which dozens of scholars have contributed with zest and perseverance. It could not have done so even if this had been the author's intention, which with al-Ghazali's sceptical attack on the philosophers' claim to indubitably certain knowledge was not really the case.

Even so, these two attempts at a general explanation of decline do provide us with a starting point for finding out why decline set in specifically around 1050. The fall of Baghdad may then be taken as symbolic of a whole series of large-scale invasions, of which the Mongol was the most ruinous but not the first. Between 1050 and 1300 enormous destruction was inflicted by a range of nomadic or semi-nomadic peoples who invaded from the pastures and deserts to the north, east and south of the Islamic world: Berbers, Mongols, Banu Hilal and Seljuk Turks, as well as European crusaders. No finely tuned civilisation can survive such attacks and plunder unscathed:

> Thus the Islam of 1300 was very different from that of 1000. The free, tolerant, inquiring and 'open' society of Omayyad, Abbasid and Fatimid days had given place, under the impact of devastating barbarian invasions and economic decline, to a narrow, rigid and 'closed' society.

And so Islamic civilisation turns in on itself and seeks salvation in a return to traditional verities which provide structure in a dangerous world that has been turned upside-down.

There is nothing specifically Islamic about this course of events. In 1241 the armies of Hulagu's cousin Batu Khan advanced from Russia to the west but at the news that his uncle, the Great Khan, had died they turned on their tracks to take part in the election of a successor. Now suppose that events had taken a different turn. If Batu's invincible warriors had continued to gallop westward because, for instance, his uncle's liver had been better able to cope with the excessive quantities of wine that he consumed, or because the clan had maintained different rules of succession, the emergent, outward-looking European civilisation might well have been laid waste and have turned in on itself in similar fashion.

The fate that Europe avoided by chance struck Islamic civilisation at its heart. There was a turning inwards directed at individual redemption and a communal emphasis on the spiritual values revealed in the literal text of the Holy Book. In many places *madrasa*s were founded, schools offering advanced courses of study in Islamic learning of the type that still exists today. 'Foreign' knowledge hardly featured in the curriculum. Peaceful coexistence gave way to a widespread belief that 'foreign' knowledge was utterly superfluous in the light of 'Arabic' knowledge, and even that the pursuit of Greek philosophy should be regarded as sacrilegious.

Such imputations had been made earlier. In Baghdad, in the heyday of the Abbasids, a certain Ibn Qutayba had hit out at foreign learning. It was flaunted even by minor officials even though it did nothing to improve people's daily lives and also distracted the young from serious study of the Koran. At the time such objections had no effect. But now in the eleventh century they struck a raw nerve. Al-Ghazali's sceptical critique was taken as a wholesale rejection of all such knowledge. And when a new mosque had to be oriented correctly, it was the scriptural scholars, the imams, who decided where Mecca lay. The subtleties of advanced spherical trigonometry were not their concern.

For centuries, the pursuit of nature-knowledge in the Islamic world was virtually abandoned.

But not forever.

In the Greek case, we cannot speak of a 'revival' though there were those rare, individual 'afterburn' effects which we discussed earlier. But in the Islamic world, such after-effects did occur in the form of regional revivals. Their plunder business completed, new dynasties arose in three separate regions, and in each there were princes who had grounds to find Greek nature-knowledge a valuable asset.

One of them was Hulagu Khan himself. In order to predict the future more accurately from the position of the celestial bodies he established the first observatory in his capital city of Maraghah. To run it he appointed a political turncoat who became one of the greatest astronomers of the Islamic world, Nasir ed-Din al-Tusi. With a few other individuals who were more or less loosely connected to the observatory, he set about revising variables and other aspects of Ptolemy's models even more radically than had been done during the Golden Age. Nowadays it is the so-called Tusi couple which has attracted most attention, a clever way of combining two circular motions to obtain motion in a straight line. Nor did revision stop there. Imagine that you were to rise from a stationary Earth up to the sphere of the fixed stars. According to Ptolemy's model of the universe, the 'wandering stars' that you would encounter in succession would be the Moon, Mercury, Venus, the Sun, Mars, Jupiter, Saturn. By contrast, al-Tusi's assistant al-Urdi produced a reasoned argument that the order of the two inner planets, Mercury and Venus, should be reversed.

In short, the traditional pattern of a Golden Age persisted, of ever bolder refinement and enrichment within the current knowledge structure. This also applied to mathematics, where Omar Khayyam (yes, indeed, the famous poet) solved cubic equations (algebra) by the intersection of specific conic sections (geometry). In this he foreshadowed a more fundamental development in Europe at the end of the sixteenth century – the rapidly increasing *identification* of algebra and geometry that was to culminate half a century later in the calculus.

Another regional revival took place in the fast-growing territory that the Ottoman sultans brought under subjugation. In their empire, nature-knowledge took the trusted form of commentaries on products of the Golden Age. For instance, al-Biruni, more Alexandrian in this regard than Ptolemy himself, had refused to follow Ptolemy's example of using natural philosophy to back up some otherwise strictly mathematical argument. In the Ottoman Empire a mathematical astronomer by the name of al-Qushji combined this standpoint with an idea that astronomers were debating at the time. This was that no observations actually ruled out the possibility of an Earth spinning on its axis. But if everything is exactly the same on a spinning Earth as on a stationary Earth, and if you further refuse to allow any natural-philosophical reasoning to settle the matter, then nothing prevents you from accepting that the Earth does indeed spin on its axis and exploring how far al-Qushji's supposition might get you in your construction of planetary models, which indeed was his main interest. And yet he never took that step. Something similar happened to al-Birjandi. He remarked *en passant* that moving objects here on Earth tend to persist on a circular path. Just over a century later, Galileo would make this idea the starting point for his radical revision of Alexandrian nature-knowledge. Al-Birjandi however left it as a passing assertion and took it no further.

In Andalusia, the third region to witness a revival of transplanted nature-knowledge, the crystallisation point was again provided by princely patronage, as it had been under Hulagu Khan and the Ottoman Sultans. Two successive dynasties of Berber princes hoped to derive benefit from promoting nature-knowledge. In this they adopted a fairly selective approach. Most of the mathematical texts were available in Andalusia, but had been put to one side because the Berber princes sought legitimacy primarily through philosophy, especially that of Aristotle. The philosopher and judge, Ibn Rushd (Averroes), exhausted himself producing commentaries on Aristotle's writings with the objective of freeing them from the Platonic gloss that al-Farabi and Ibn Sina had given them on the other, eastern side of the Islamic world. This rather purist approach to Aristotle was also maintained in other respects. Seldom has the

distinction between 'Athens' and 'Alexandria' been expressed so cat-
egorically as by Ibn Rushd. He was familiar with planetary model
building in Ptolemy's vein, yet would have nothing to do with it:

> The astronomical science of our days surely offers nothing from
> which one can derive an existing reality. The model that has
> been developed in the times in which we live accords with the
> computations, not with existence.

Ibn Rushd also defended the indisputable truth of Aristotelian doctrine against al-Ghazali's sceptical objections. Amazingly, this has led
him to be hailed as an early Voltaire of sorts by some modern-day
partisans of the Enlightenment, the very movement that so ardently
advocated the fundamentally open character of all knowledge against
its dogmatic closure!

The rise and decline of nature-knowledge in Islamic civilisation,
then, follow roughly the same pre-modern pattern as with the Greeks.
In both cases the downturn following the Golden Age is sudden and
steep, and distinguished by a few individual late-comers of remarkable
ingenuity. Except that in the Islamic world these late-comers are
firmly anchored in three episodes of regional revival: Persia under
the Mongols; the eastern Mediterranean under the Ottoman Turks;
and Andalusia under the Berbers. And it is the gradual defeat of the
Berber princes by Christian Spain that creates the conditions in which
the second transplantation could take place.

MEDIEVAL EUROPE: SPECIAL DEVELOPMENTS

The fact that Gerard of Cremona and his colleagues drew on texts in
Andalusia had vast consequences for events in Europe. Greek nature-
knowledge had developed rather differently there than in the eastern
Islamic world. In contrast to Baghdad under the Abbasids and later on
to Renaissance Europe, medieval nature-knowledge was almost exclusively oriented towards Aristotelian doctrine. What little knowledge
European scholars possessed of other Athenian philosophers fell away
after the completion of the huge translation project which Gerard and

his colleagues undertook in Toledo, leaving Aristotle with a virtual monopoly. Dante referred to him characteristically in the *Divina Commedia* as 'the master of those who know'.

His supremacy extended not just over philosophy but also over mathematical nature-knowledge. Gerard of Cremona certainly translated core Alexandrian texts such as Ptolemy's *Almagest* and Euclid's *Elements* from Arabic into Latin. But their content was subsequently made more accessible only in simplified form. Moreover, their mathematical proofs were converted into the logical forms of argument that mark Aristotelian doctrine. Earlier I have made a point of the contrast and lack of interaction between 'Alexandria' and 'Athens'. In Andalusia, and even more so in medieval Europe, this degenerated into a wide-ranging subjugation of the mathematical mode of nature-knowledge to a single representative of its natural-philosophical counterpart.

This Aristotelian monopoly was furthered not only by the one-sidedness of the Andalusian text corpus but also by the circumstance that the translation of the full collection of Aristotle's writings into Latin coincided with the founding of the first universities. Every student, before completing his training as a physician, lawyer or cleric, had first to follow a curriculum in the 'arts faculty' ('arts' in the sense of the various disciplines on offer in the curriculum). Aristotle's teachings were made more easily digestible by means of extensive commentaries and in that form proved well suited as a general introductory course. Between *c.* 1250 and *c.* 1650, there were very few academically educated Europeans who lacked any familiarity with at least some of that knowledge.

At first these introductions to Aristotle's ideas were taught almost exclusively by monks and priests, which might suggest that their acceptance had caused little if any difficulty. But that is far from being the case. Many of Aristotle's ideas do not appear to sit at all well with Christian doctrine. A Christian believes that after death the soul leaves the body, whereas Aristotle considered the soul to be a form indivisible from the matter which it animates. For Aristotle the world is eternal, whereas a Christian sees time as sandwiched between the

Creation and the Last Judgement. But in the first half of the thirteenth century a Dominican monk named Albert and his pupil Thomas from Aquino became excited by the new perspectives on nature that had been opened up by the recently translated works of Aristotle, while being only too aware of the theological pitfalls. It seemed to them to be worth trying, in so far as this could be done at all, to remove those obstacles. But how? Bulky volumes produced by Albertus Magnus (the Great) and Thomas Aquinas and known as *Summae*, all-embracing summaries, explicated and where necessary adapted Aristotelian doctrine. In this way the teachings of Christianity became closely entwined with the philosophy of Aristotle.

What made the accomplishment possible in the first place was Thomas' acute recognition of the core problem, and his ability to come up with an ingenious solution. That core problem lay in God's sovereign will. According to the Bible, God has designed Creation as He pleases and He is at liberty to do with his creatures as He pleases. In Aristotle's view, the world cannot possibly be any different from what it is; it is a world of necessity, not of freedom. Thomas brilliantly perceived that God's omnipotence lends itself to being reconciled with the limitations that Aristotle had imposed on any creator. He did so by distinguishing (medieval scholars were masters in making distinctions) between God's absolute power and His 'preordained' power. In effect, this meant that, although God is perfectly capable of intervening in the natural order whenever He wants, He has of His own free will decided not to do so. He may be able to will it, yet He is under no self-imposed obligation to will it.

Void space offers a good illustration of how Albert and Thomas gave direction to later medieval commentaries on Aristotle. In nature we never come across a vacuum, and Aristotle had shown logically that there can be no such thing. If you drop an object into water or syrup it falls more slowly than through the air and that observation leads on to the generalisation that the rarer the intermediate substance is, the faster the object will fall. From this follows that if the substance were infinitely rare, the speed of the object would likewise become infinite.

But that is not possible, as an object cannot be in two places at one and the same time. Therefore no space can be empty – a vacuum cannot exist. That was Aristotle's reasoning and it makes perfect sense (except that ever since Galileo we know that the generalisation cannot be allowed). Now what does this mean for God? If He is unable to create a vacuum, this inability of His inevitably places limits on His sovereign will. If we look at the lecture notes of one of the most innovative commentators on Aristotle, Jean Buridan, we see that he deals with the question of the vacuum in two parts. First of all he argues that God, if He should so wish (though you may be confident that as a rule He will not), could create an empty space. Buridan even suggests various ways in which God might set about it. But in the following, much fuller section he then relaxes and goes on to defend at great length Aristotle's proof of the impossibility of the vacuum against the many objections that (by way of an exercise in coherent reasoning) earlier commentators had raised against its compelling power.

In this way the Aristotelian monopoly was established, with profound consequences. There was certainly mathematical talent around that was seeking an outlet, but since it could not be found in an Alexandrian direction it had to take the few quantitative opportunities offered by Aristotle. In Oxford a group known as the Calculators was formed who wherever it seemed feasible adopted a mathematical approach to philosophical problems. Aristotle had been clear that an object's quality can never be reduced to something quantitative (in the Aristotelian conception of things the modern definition of the colour red as a wavelength of 0.0008 mm would not even begin to make sense; it would be literally meaningless). But the Calculators hit upon the idea that we can attribute greater or lesser intensity to qualities. One particular flower, for instance, might be a brighter red than another. And with those intensities and their uniform or non-uniform rate of change you can carry out a large variety of at times quite involved calculations. In Paris, Buridan's pupil Nicole Oresme picked up on the idea and imaginatively made it visible by way of a graphical representation.

The same Oresme also toyed with the idea of the Earth rotating on its axis. In an elaborate argument he presented both the philosophical and the empirical arguments for it, only to reject it in the end as no more than an intellectual game which would in any case conflict with certain passages of Scripture. In short, even the most innovative thinking about nature in the period between 1250 and 1450 remained confined within the accepted (though inventively stretched) conceptual framework.

With just a few exceptions. Just as the Islamic world's Holy Book threw up the tricky mathematical problem of calculating the direction of Mecca when building a mosque, so Christian Europe faced the problem of calculating the changing date of Easter. Working within the preconditions laid down by the Gospels on the one hand, and the Julian calendar on the other, skilful mathematicians eventually succeeded in working out a satisfactory solution. Unlike the building of new mosques, that solution was faithfully adhered to by the Church.

It was not only the dating of Easter that (inevitably) lay beyond the scope of Greek nature-knowledge; it was also true of a couple of other medieval treatises of quite a different nature, which dealt with the subjects of falconry and the working of magnets. The author of the first was Emperor Friedrich II of Hohenstaufen, then as now known as *stupor mundi*, the wonder of the world. He was always out to get to the bottom of things. His treatise on 'The Art of Hunting with Birds' breathes an 'intense curiosity about the particulars of nature, most unusual in an age that was forever seeking universals'. To find out whether vultures find carrion by sight or smell he had the eyelids of a vulture sewn up: it completely ignored the carrion that was offered. Moreover, they do not eat anything but carrion. Vultures that had been deliberately starved were still not interested when young chicks were offered to them. Friedrich's curiosity always had a practical objective – his careful observations and description of the behaviour and habits of different birds were intended to improve falconry.

In the Middle Ages the only other example of such an empirical, practice-oriented approach to natural phenomena is Pierre de Maricourt's treatise on magnets. He noted, for instance, that however often you divide a magnet in two each piece retains a north and a south pole. And he described in careful detail how a magnetised needle behaves when you let it revolve freely. His ultimate objective was to understand and improve the workings of the compass in the interests of navigation.

	Greece	Islamic civilisation	Medieval Europe
UPSWING	creation + (in mathematics) transformation	translation ⇒ enrichment	translation ⇒ *Aristotelian* enrichment
• **Golden Age**	Plato to Hipparchus	al-Kindi to Ibn Sina, al-Biruni and Ibn al-Haytham	Albert the Great to Oresme
DOWNTURN	*c.* 150 BC	*c.* 1050 AD (AH 440)	*c.* 1380
(1) Why?	'normal'	'normal'	'normal'
(2) Why at that time?	sceptical crisis? cessation of patronage?	invasions ⇒ turning inwards	*possibilities for enrichment exhausted*
(3) Aftermath	steep, protracted; generally low level	steep, protracted; followed by partial, higher-level reversal	steep, protracted; *unreversed*
• **marked by**	codification, commentary, syncretism ...	commentary mostly	commentary
• **still outstanding**	Ptolemy, Diophantus; Proclus, Philoponus	Nasir ed-Din al-Tusi, Ibn as-Shatir, Ibn Rushd	*none*

Medieval nature-knowledge, too, reveals an unmistakeable pattern of upswing and downturn. Once again historical comparison can help us to grasp its chief characteristics. In the table they are shown in italics.

We have spent long enough on the most obvious medieval characteristic: namely that of the five Athenian schools only Aristotle's remained, and that 'Alexandria' either went underground or was rephrased and trapped within an Aristotelian mindset. I also noted that this Aristotelian monopoly did not stand in the way of another upswing – the one associated with the names of Albert the Great and Thomas Aquinas, and which culminated in another Golden Age. Once again we encounter in the space of a few generations a constellation of innovators of high calibre, notably Buridan, Oresme and the Oxford Calculators. But again the Golden Age ends abruptly, and with the death of Oresme in 1382 it is effectively over. Only this time the decline is steeper and even more definitive than before; there are no great 'late-comers'. Buridan's concept of 'impetus', the conclusions of the Calculators and Oresme's graphic representations are mulled over by later commentators with but little variation and no addition of anything new. It is this fifteenth-century period that gives 'scholasticism' its enduring reputation for pointless hair-splitting. In the history of nature-knowledge nothing will ever again be quite so dry as dust and sterile as those late medieval treatises. This particular difference, too, is due to what, ever since the Toledan translation effort, had made medieval nature-knowledge so one-sided. If thought remains encapsulated within a single system because there is no available alternative, even a fairly flexible system such as Aristotle's will in the end become locked up within itself. Attempts at innovation soon hit the limits of its boundaries and eventually run into the sands of dogmatic pettifoggery. Only when there is competition can the need for dialogue arise, and boundaries may have to be extended. It was the fall of Constantinople in 1453 that opened the door to just such competition.

RENAISSANCE EUROPE: SPECIAL DEVELOPMENTS

After 1453, Italian scholars – spurred on in the first place by Cardinal Bessarion – began to translate one original Greek work after another into Latin. The translation movement spread rapidly through the whole of Europe, and within a century and a half the classical spectrum of all five Athenian schools plus the Alexandrian mode of nature-knowledge was fully restored. Those who brought this about are usually called 'humanists'.

'Humanism' is a concept with quite a range of current meanings. Here I take it to stand for a movement animated by an urgent desire to renew knowledge by returning to classical antiquity across an intervening 'middle' age now perceived as dark and sterile. Turning back to the original sources of classical civilisation took many forms and also embraced similar efforts in painting, literature and music. Building on the recovery of ancient texts, the humanists also cherished a pedagogic ideal, though it was not tied as such to any specific philosophical school. Sceptical mockery set the tone in many of the writings of well-known humanists such as Erasmus and Thomas More who drew heavily on the arsenal of anti-philosophical arguments developed by Pyrrho and the later Sceptics. But most humanists tended to follow one of the dogmatic Athenian schools or to focus their attention on mathematical texts. Not only was Aristotle himself now translated directly from Greek into Latin, but so were Plato's dialogues, the surviving texts of the atomists and the Stoa, and virtually everything that we now possess of Alexandrian writings. The classic philosophical debates revived and again took the form of exchanges of propositions derived from first principles that proponents considered to be beyond doubt. As before, such exchanges were not particularly fruitful. True, they were now enlivened by the rise of the printing press, but this innovation did nothing to change the nature of the debate; it only enabled a larger number of people to become aware of it and to take part. Meanwhile, in Europe's universities teachers of Aristotle's doctrines perceived the loss of their monopoly as a challenge. They expurgated available texts by spotting and removing errors that had crept in during the process of translating them from Greek into Syriac, from Greek or Syriac into Arabic and then from Arabic into Latin. And they modified their style of teaching to accord with the more flexible didactic methods favoured and devised by some humanists.

The most radical change brought about by humanists in the domain of nature-knowledge occurs in the Alexandrian tradition. Take an oration which Regiomontanus, a protégé of Cardinal Bessarion, gave in Padua in 1464 in which he eulogises mathematics and mathematical nature-knowledge:

> The theorems of Euclid have the same certainty today as a
> thousand years ago. The discoveries of Archimedes will instill no
> less admiration in men to come after a thousand centuries than
> the delight instilled by our own reading.

Regiomontanus followed Ptolemy in contrasting the certainty of mathematical knowledge with the fake certainties with which philosophers were in the habit of assaulting each other. Their higher social standing, he argued, was misconceived and the much larger salaries that they were paid by the universities were quite unjustified. It was, he argued, high time that mathematical nature-knowledge was accorded the respect it deserved.

In his oration, Regiomontanus also proposed an elaborate programme for publishing ancient mathematical texts, and eight years later, having acquired his own printing press, he set about doing just that. Four years later, aged 40, he died, but the task that he had begun was taken over by others. By 1600 most of his publishing programme had been completed though his social ambitions for mathematicians remained as distant as ever. As a rule, the mathematician remained, or rather became (since the Alexandrian mode of nature-knowledge had barely been recognised during the Middle Ages) a courtier subject to the fickleness of patronage. Occasionally he might be a professor instead, but cold-shouldered by his philosophical colleagues and earning less than a quarter of what they were being paid.

As for the actual content of mathematical nature-knowledge, it was not only a question of translation and study but also to some extent of enrichment. Yet once again it stayed within the traditional framework: there is no increase beyond the number of five original subjects and far-going abstraction remains the norm. An exception to the rule, but only in retrospect and more apparent than real, was a book published in 1543, the year when the author died – it is not even certain whether the publisher was able to send him a copy in time. The Latin book bore the title 'On the Revolutions of the Celestial Spheres' and the author was a Polish-German canon named Nicolaus Copernicus. His objective was, in the spirit of Regiomontanus, to restore astronomical knowledge to its ancient purity. In this he closely followed the arrangement and the approach of Ptolemy's *Almagest*, but with three exceptions.

In an introductory chapter, Copernicus explains that if you consider Ptolemy's mathematical models of the heavenly bodies, not individually but as a unified system, it has in the course of time turned into a 'monster'. Head, torso, arms and legs, crooked in respect to each other, are no longer in proportion. It is a shortcoming that cannot be rectified by tinkering with it on a limited scale; a more radical overhaul is required.

To embark on such an enterprise was all the more necessary, Copernicus argued, because one of the three devices that Ptolemy used to represent the irregular movements of the planets by means of regular circles is really a misfit. The so-called equant point, albeit technically unexceptionable, owed its introduction to what was in effect no more than a sleight of hand to obey at least formally the basic requirement of uniform angular motion. Al-Tusi had raised the same objection to Ptolemy's equant point, and solved it with the 'Tusi couple', which Copernicus now also used, almost certainly independently. Copernicus definitely remained faithful to the prevailing conception of the mathematical astronomer's task – his job was, and remained, to 'save' the phenomena. But now, with the self-inflicted loss of the equant, he was left with only two mathematical tools at his disposal. Here is another reason why the wished-for return to the purity of the original could be achieved only by means of a full-scale revision.

What, then, could serve as the starting point for such a revision? Looking around classical antiquity, he encountered an idea that might possibly be suitable: Aristarchus' notion that the Sun, not the Earth, should be regarded as the centre of the universe. But what for Aristarchus was just an idea, Copernicus sat down to calculate through from beginning to end in the style of Ptolemy.

Thus Copernicus was hardly the revolutionary figure that he looks like when viewed in retrospect. He was not out to create something new, but to restore the old, using tools which were also drawn from antiquity. That was also how his contemporaries saw him and his work, and for over half a century until about 1600 that would not change.

In Books II to VI they found the familiar kind of model building as in the *Almagest*, except that it now lacked the equant point and was oriented towards the Sun instead of the Earth, which now appeared as another planet. Furthermore, Copernicus had incorporated a number of improved observations. As in the *Almagest* his readers found abstract models, suitable for predicting the future position of the heavenly bodies but not for representing their true trajectories. And whoever followed the calculations in Books II to VI encountered along the way dozens and dozens of auxiliary circles (even more than with Ptolemy) which it was impossible to imagine could actually be present in reality.

It was only in Book I that Copernicus presented his full model, but now in greatly simplified form, as if each heavenly body traces one circle in uniform motion. And there he argued that the Earth is not just in his abstract model but also in reality a planet that spins on its axis once every day and revolves around a stationary Sun once every year. He was aware that it would be regarded as quite an odd assertion; he even hid the finished work away in a drawer for decades for fear of being ridiculed should he ever publish it.

Even so he did back up the case with some arguments. He could not take these from observation since there were as yet no observations to indicate a moving Earth. He had to rely on a range of considerations taken from natural philosophy (much as Ptolemy had done earlier, though naturally different considerations). To his contemporaries they looked quite as unconvincing as, in retrospect, they look to us. They were certainly not capable of refuting the obvious objections to a rotating Earth: we *see* the Sun rise and set; we cannot *feel* the Earth turning; clouds and birds would surely be left behind if it did. For the rest, Copernicus made much of the simplicity that results from placing the Sun at the centre. The simplification is fairly technical but means that some data that in Ptolemy are just incidental facts appear in Copernicus as necessarily thus and not otherwise. This affected in particular the true order of the planets (at issue, too, in view of then-current astrological debates). Placing the Sun at the centre unambiguously fixes the order of the planets. In Copernicus' system, the observational data leave no possibility other than that, working outward from the Sun, Mercury is the first interior planet and Venus the second.

Running through Copernicus' work, then, is a conspicuous fault line. Book I with its simplified presentation, its argument taken from bits and pieces of natural philosophy and its somewhat convoluted appeal to enhanced inner coherence was hard to bring in line with Books II to VI with their complicated model building on Ptolemaic lines. In our time Books II to VI are largely ignored and attention is focused on the seemingly modern Book I. In the later sixteenth century it was precisely the opposite. Mathematical astronomers eagerly made use of Copernicus' extensive planetary models to draw up their tables and almanacs, with little concern for the absurd claims in Book I. Until 1600 no more than eleven people had spoken out in favour of Copernicus' revision. That low level of support had nothing to do with religious objections from churchmen, which amounted at most to some scattered grumbling. Moreover, what little acceptance there was turned out to be fairly selective, in that, for instance, some accepted the Earth spinning on its axis but not its revolving around the Sun. And nine of the eleven did not even regard the matter as of any great, let alone decisive, importance. They accepted the 'opinion' of Copernicus without drawing possible consequences for their other ideas and conceptions. Seen in the context of the time, Copernicus' book fitted seamlessly into the pattern that I have sketched in this chapter: enrichment of the abstract-mathematical mode of nature-knowledge while preserving the traditional knowledge structure. However, we can now see that buried in the cellars of that technical-astronomical treatise there lay a ticking time bomb. But it was far from certain that it would ever be detonated.

So much for the fortunes in Renaissance Europe of our old acquaint-ances, 'Athens' and 'Alexandria'. But more happened at the time than the revival and enrichment of these two only. What we saw earlier in incidental fashion with Friedrich II and Pierre de Maricourt takes wing after the fall of Byzantium. Halfway through the fifteenth century, on the margins of reviving Greek knowledge, appeared a third mode of nature-knowledge that was quite different from the two Greek modes. Its adherents derived their verities not from the intellect but from accurate observation, usually with a practical objective in mind.

Some focused their efforts on describing things with extreme care and accuracy. A well-known example is Vesalius with his mag-nificent anatomical atlas in which the descriptions, but especially the engravings, revealed the human body with unprecedented accuracy.

But also in many other fields there was a desire to avoid hasty generalisation or sweeping explanations based on a minimum of fac-tual material. Phenomena had first to be mapped out with patience and dedication. This was done quite literally in the form of world maps and atlases, while private collectors drew up detailed catalogues of what were the earliest museums. These contained collections of the most diverse objects and their owners were striving all the time to put the whole world, as it were, into display cabinets.

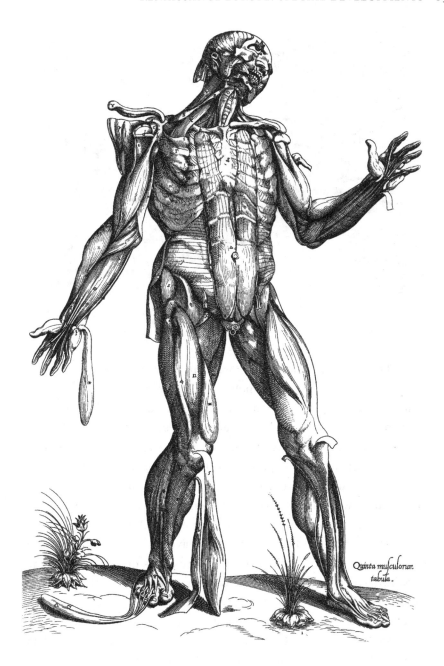

*Quinta musculorum.
tabula.*

The heavens were also mapped out with unprecedented precision. Tycho Brahe, a Danish nobleman, noticed as a young student that a conjunction of Jupiter and Saturn occurred a good month earlier than the almanac predicted but that nobody seemed concerned about it. That, he felt, would have to change. He mastered the discipline of astronomy and eventually persuaded the king to grant him the use of a little island in the Sound on which to build an observatory. He equipped it with a range of instruments which he himself either invented or greatly improved, and named it 'Uraniborg' (heavenly fortress). By the time he was forced to leave his island a quarter of a century later, he had pushed the accuracy of his observations to the very limits of what was attainable with the naked eye. While disagreeing with Copernicus' idea of the structure of the universe, he did feel that a definitive theory about it, and about the irregularities in planetary movement, could be built only on the basis of the mass of data that he had so painstakingly collected during his nocturnal labours.

Dürer's perspective grid

Precise description was the order of the day in other fields as well. In Germany, three scholars in succession published herbals with more or less true-to-life engravings and with detailed verbal descriptions of the appearance and properties of a large variety of herbs. In Goa, a Portuguese colony in India, the apothecary Garcia de Orta did the same thing. A practical objective was clearly implicit in every case, since the herbs described were primarily those to which, rightly or wrongly, healing properties were attributed.

In other cases practical use was quite explicitly the primary purpose, and now for the first time mathematics proved capable of useful application. A classic example is linear perspective. Painters of the Italian Renaissance were increasingly keen to break free from the static Byzantine mode of painting and adopt a more natural, true-to-life style. In this quest they invented perspective and called on geometers to help them calculate precisely the vanishing point and convergent lines.

The geometry needed for that was already to hand, more or less. The novel element was that it was now no longer cultivated for its own sake. Geometers also lent a hand in gunnery, in fortification and in calculating geographical location. In the Netherlands, Simon Stevin even applied his mathematical knowledge to improving the performance of windmills. His proposals, though well thought through, met with little success. He registered patents on a whole range of improvements but it was precisely those that he had developed through mathematics that had little effect. And indeed that is how it usually went. At this stage, mathematical techniques only made a real difference in perspective, fortification and navigation. Most often craftsmen continued to rely on their tried and trusted rules of thumb without resorting to any theory. The successful application of theory to practice was far too random and incidental to speak of any breakthrough to our modern, mathematically underpinned technology.

It was not only individual mathematicians like Stevin who undertook a quest for nature-knowledge of a kind that should deliver practical results. Craftsmen, especially the more creative ones, hoped it would enable them to develop new or greatly improved tools. The prime example is Leonardo da Vinci. He was one of those painters who strove for true-to-life representation. But he was also a naturalist, as well as being a craftsman extraordinarily at home with the construction of instruments and tools. What makes him unique is that with him all these activities flowed into each other. He investigated and drew with great precision how muscles and tendons are attached to the bones in birds, and subsequently tried to imitate it with a system of pulleys that he hoped would enable humans to fly. It is known for sure that he actually undertook trial flights, with readily predictable results. He was also interested in optimising the performance of mechanical appliances. He realised that to this end friction has to be reduced to a minimum and conducted experiments to find out on what specific factors friction depends. That way he discovered that the size of the contact area does not matter but that friction increases in proportion to the load.

Finally, much of this research with a practical purpose was also steeped in magic. That was particularly the case with Paracelsus. His name was actually Theophrastus Bombastus von Hohenheim (the word 'bombast' is derived from it) and he was an alchemist who argued, in language as fiery as it was muddled, for a radical break with the classical tradition. He continued the quest undertaken in

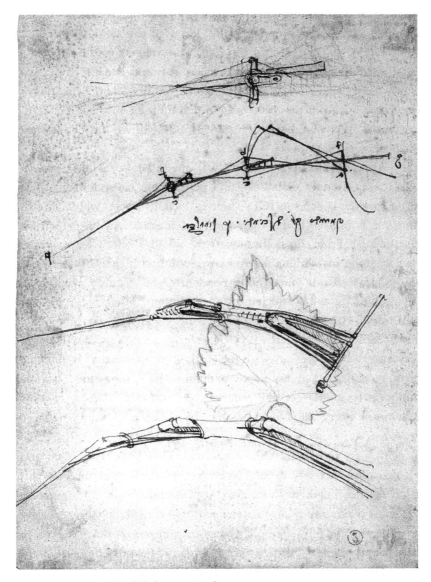

Leonardo: Bird flight examined
In the lower half are two sketches of a tendon used to open and fold the
wing; above these are two sketches of how to imitate this by means of a
system of pulleys. As usual with Leonardo, the text is in mirror writing; it
reads 'quando g discende p sinalza' ('when g descends, p rises').

Islam for the 'elixir', the means of speeding up the process of change in metals as well as prolonging life. He used the same techniques: combustion, distillation and the like. The imagery with which the great alchemical work, the purification of metal and the soul, was expressed was equally esoteric, though naturally it did not draw from the Koran but from the Bible. And, much more than in Islamic civilisation, all this was deeply embedded in a detailed, fully developed magical world picture.

At that time a distinction was made between black magic, which was literally diabolical, and permissible natural magic. The latter saw the cosmos as full of hidden, 'occult' forces, which though invisible could, with the right forms of incantation, be tapped for personal use. From this point of view the world is a network of objects that in some mysterious fashion correspond with each other. When things possess this inner correspondence, the occult forces make themselves known through mutual attraction; otherwise repulsion will occur. Not only is the cosmos a network, it is also animated, and according to Paracelsus it further constitutes, as it were, a chemical laboratory. In analogy with the Holy Trinity all matter is made up of three basic principles, 'sophic sulphur' (the combustible), 'sophic mercury' (the volatile) and 'sophic salt' (solidity). The cosmos is reflected in every individual human being and deep within us the same processes are at work in miniature. We are healthy when the three principles are in balance with each other; in sickness Paracelsus and his disciples are ready to feed us certain minerals to restore the balance. In this, chemistry and alchemy ('chymistry' in contemporary parlance) overlap to a large extent, and its practitioners are generally apothecaries and physicians. This medical component led to the whole field of activity being named 'iatrochemistry', healing chemistry.

This 'third' mode of nature-knowledge that appeared in Renaissance Europe from out of almost nowhere had quite a different attitude to classical antiquity from that taken by those humanists who were occupied with the revival and possible enrichment of Athenian natural philosophy and Alexandrian mathematical nature-knowledge.

Many attacked classical wisdom as such (though they were much more deeply imbued with it than they realised). Petrus Ramus contemptuously rejected Euclid's mathematics, arguing that real mathematics was to be found on the streets and in the marketplace where actual people were actually calculating. The potter, Bernard Palissy, who collected fossils and realised that they were petrified creatures and plants, praised his own collection of natural objects on the grounds that experience is forever the mistress of the arts. A two-hour visit to his collection, he claimed, would teach you more than forty years of studying ancient texts. Paracelsus went even further and created a scandal by publicly burning an authoritative textbook on Galenic medicine.

It is not surprising that those who pursued this path to nature-knowledge were generally not the same people as the natural philosophers and mathematicians. Much as 'Athens' had stood apart from 'Alexandria' and there had been little exchange between them except for some rather unconstructive polemic, so there was a virtually watertight separation in respect of this 'third' mode of nature-knowledge.

To this general rule of watertight separation there are two apparent exceptions. One was a scholar who not only operated along 'Athenian' lines but also in empirical and practical fashion. In around 1538, the Portuguese humanist and explorer João de Castro wrote a treatise which was virtually indistinguishable from countless medieval Aristotelian works except for a single correction – the voyages of discovery made by his compatriots enabled him to criticise in passing Aristotle's assertion that on Earth there is much more land than sea. Meanwhile the log books which he kept of his own sea voyages to and from India breathe a very different atmosphere. Here it is not the dogmatic repetition of traditional verities that sets the tone but rather open-minded, unprejudiced observation. Here, Castro is not the philosopher but an empiricist seeking to get a grip on natural phenomena in all their capricious unpredictability. How about those factors which are so essential to navigation – wind, water and the means of charting a course when land is out of sight?

He interrogated Ethiopean chieftains about the sources of the Nile. He noted down the odd behaviour of compass needles in a variety of circumstances until he realised that proximity to iron objects seriously disrupts their working. His systematic approach to measurement did not prevent him from being extremely cautious about the results, so if any doubt remained he would ask others to make the same measurements. Indeed, this future viceroy of Portuguese India would go so far as to hand the Jacob's Staff to the bosun or even a simple caulker and take their recorded results as seriously as those he had made himself.

There is another apparent exception to the rule of strict separation between the three modes of nature-knowledge that flourished during the Renaissance. Among the Aristotelian humanists there were some who realised that since their medieval monopoly had been broken the texts had to be expurgated and their teaching had to be adapted to the new times. These men also did their best to enrich Aristotelian doctrine by turning to its two principal rivals, the Alexandrian and the empirical-practical modes of nature-knowledge.

In the latter the magical aspect was paramount. Some scholars tried to take the central idea of occult forces working between heaven and Earth, and of the hidden correspondences through which they operate, up in Aristotle's doctrine of substance and form. So the philosopher-physician Jean Fernel argued that under the influence of occult forces the form and, with it, the essence of our body can be assaulted by certain pathogens. Epilepsy or an outbreak of the plague, he argued, cannot possibly be explained in the ordinary manner as due to an imbalance of fluids or minerals in our body. What actually happens in his view is that its essential form comes under attack. And now this thoroughly Aristotelian philosopher has set the stage to a sufficient degree to subject the efficacy of various remedies to patient empirical testing.

Similarly Christoph Clavius, an early member of the Jesuit Order, attempted to come to some accord with the competition without sacrificing what was proper to the 'Athenian' knowledge structure.

He did so by opening up Aristotelian doctrine in the other, Alexandrian, direction. He produced a Latin edition of Euclid's *Elements* in which he represented the strictly mathematical form of proof as equal in value to Aristotle's style of reasoning. Thereby, almost singlehandedly, he created an intermediate intellectual discipline of 'mixed mathematics' in which calculations and Aristotelian explication went hand in hand. In this he went a great deal further than the medieval Oxford Calculators. Anything observable that contained any kind of quantitative element was treated by Clavius in arithmetic or geometric style but at the same time in a philosophically acceptable manner. His efforts ranged from sundials and the calendar to planetary trajectories and the nature of light and vision. His name is closely associated with the Gregorian reform of the calendar of 1582 in which ten days were dropped to bring church holidays back in line with the seasons.

When Clavius began his quantitative enrichment of Aristotelian doctrine in the mid sixteenth century he found no support among his fellow Jesuits. His campaign to have 'mixed mathematics' adopted in the prescribed educational programme met with indifference. By his death in 1612 things had changed a great deal. Not only was mixed mathematics now part of the curriculum, but he had also built up a team of like-minded scholars at the Collegio Romano, the central Jesuit university, who were chiefly engaged in astronomy. He and some of his younger colleagues viewed with sympathy, and even with a really misplaced sense of affinity, the innovative work which two scholars of a typically Alexandrian orientation were introducing to the world of scholarship in the early decades of the seventeenth century. Their names, which we shall encounter again in this book, are Johannes Kepler and Galileo Galilei.

A TREND-WATCHER IN 1600: THE THREE TRANSPLANTATIONS COMPARED

As we now know but nobody then could have known, we have reached the point when a revolution was about to break out.

Revolution? Pardon me! Are not historians of science nowadays in near-unanimous agreement that what our predecessors once labelled 'The Scientific Revolution' was no such thing, but a gradual process without any obvious markings or fault lines? A leading historian of science, Steven Shapin, won much applause for his scornful reference to earlier historians who 'announced the real existence of a coherent, cataclysmic, and climactic event that fundamentally and irrevocably changed what people knew about the natural world and how they secured proper knowledge of that world'. In his view, seventeenth-century nature-knowledge breaks down into just 'a diverse array of cultural practices'.

The 'either/or' way of thinking and the choice of adjectives are characteristic: *either* modern natural science must have emerged as a single, coherent whole in a single, sudden, dramatic explosion (cataclysm = sudden, large-scale conflagration), *or* seventeenth-century nature-knowledge constituted, as in earlier and later times, a loose collection of events which, if we wish to come to historical grips with them, we had better not link up with natural science in its present-day guise.

Amidst all this forceful rhetoric there is an urgent need for a middle way. Let us resist the temptation to indulge in counter-rhetoric, and take as empirical a path as we can find in our historians' toolkit. A quintessential feature of revolutions is that their occurrence only looks obvious in retrospect. Nobody at all foresaw the fall of the Berlin Wall in 1989; at best there was a vague idea that not even communist regimes would last forever. Even in 1988 not a single journalist or newsreader even hinted at a speedy disintegration of the Soviet empire. In our case the events occurred much longer ago, but we can try playing a trick. Let us assume that at the end of the sixteenth century there was already a European Commissioner for Science Policy. Let us imagine that, to mark the approaching turn of the century, he invites a science journalist of his acquaintance to write a report. A Eurocommissioner, after all, does not distribute his subsidies at random. The reporter is instructed to compare precedents, to chart the predominant trends and then to extrapolate these trends into the future. For that is, after all, how forecasters generally go to work: by extrapolating from existing trends. Our reporter has no inkling of what will happen after 1600. He can only work with what he knows of how nature-knowledge was pursued in the past – all of it (in reality his knowledge could have been quite limited only, as by 1600 large stretches of the Middle Ages had already disappeared over the horizon and little had ever been known in Europe of pertinent activities in the Islamic world). In this manner our fictional report will provide us with a yardstick by which we may determine, not *whether* but *to what extent* continuity predominated and *to what extent* we may therefore speak of a more or less radical break with the past. The *more closely* the predictions that our reporter derives from the trends he has been watching approximate to what actually happened after 1600, the *less* valid are the grounds for calling what happened in nature-knowledge in the decades around 1600 a veritable revolution.

Let us now steal inside the commissioner's office and read the report with him, bearing in mind that the style is not so flowery as it would have been then, nor so bureaucratic as it would be nowadays:

> *Earlier periods of upswing in Greek nature-knowledge resulted in a Golden Age. The first question that I should like to ask is whether we can apply that description to the present age.*
>
> *I define 'Golden Age' as a 'strikingly dense constellation of great creative talents'. And then the answer is obvious. What expression is more appropriate to indicate a period of a century and a half in which the 'Alexandrian' legacy has been restored and enriched by scholars of the calibre of Regiomontanus, Copernicus and Stevin, and the 'Athenian' by scholars such as Clavius and Fernel? In addition, at the same time practical scholars and learned practitioners such as Leonardo, Vesalius, Tycho, Paracelsus and Castro have created a new mode of nature-knowledge which is chiefly directed at accurate observation. And this still leaves out*

	Greece	Islamic civilisation	Medieval Europe	Renaissance Europe
UPSWING	creation + (in mathematics) transformation	translation ⇒ enrichment	translation ⇒ Aristotelian enrichment	translation ⇒ enrichment
• Golden Age	Plato to Hipparchus	al-Kindi to Ibn Sina, al-Biruni and Ibn al-Haytham	Albert the Great to Oresme	Regiomontanus + Leonardo to Stevin + Vincenzo Galilei + Clavius
DOWNTURN	c. 150 BC	c. 1050 (AH 440)	c. 1380	???
(1) Why?	'normal'	'normal'	'normal'	'normal'
(2) Why at that time?	sceptical crisis? cessation of patronage?	invasions ⇒ turning inwards	possibilities for enrichment exhausted	possibilities for enrichment exhausted?
(3) Aftermath	steep, protracted; generally low level	steep, protracted; followed by partial, higher-level reversal	steep, protracted; unreversed	–
• marked by	codification, commentary, syncretism …	commentary mostly	commentary	–
• still outstanding	Ptolemy, Diophantus; Proclus, Philoponus	Nasir ed-Din al-Tusi, Ibn as-Shatir, Ibn Rushd	none	–

those many persons who during the past century and a half have practised and still practise nature-knowledge at the same high level as the ten whose names I have mentioned. As the year 1600 approaches, which after all is our deadline, I am confident that we have as much right to speak of a Golden Age of nature-knowledge in Renaissance Europe as earlier in Greece, in Islamic civilisation and in medieval Europe.

How far can we take this comparison? Do the similarities in upswing allow us to deduce anything about what might follow? Is it possible to say anything at all sensible about it? Perhaps a table showing the current upturn will help.

In the table I have already inserted one important event which we may expect to repeat itself. Each of the three previous Golden Ages ended in a downturn that was as sudden as it was steep. That is the normal course of events. It is what we should expect and no special explanation is therefore needed. So I venture to assert that there is a probability verging on certainty that our Golden Age of Renaissance nature-knowledge will come to a similar end.

That leaves the much more difficult question of when and how this decline will occur.

Let us inspect the table of the three transplantations more closely. We have to take into account all the historical information at our disposal. And what immediately draws our attention is a remarkable deviation in the medieval pattern. In contrast to the other two, transplantation in the European Middle Ages remained confined to 'Athens' and even to a single part of it, the Aristotelian. That one-sidedness had far-reaching consequences. It drastically limited the opportunities for development which cultural transplantation offers; also, what further development there was became bogged down in sterile repetition which did not even leave room for the occasional brilliant late-comer.

But more important for our purposes are the similarities between the other two. In Islamic civilisation, just like nowadays in Renaissance Europe, revival takes place over a broad front.

The forms of enrichment are remarkably similar; the Greek pattern is embroidered upon with great inventiveness and, if you did not already know, you would be hard put to guess whether some discoveries were made in Islamic civilisation or by us. Added to that, on the margins of Greek nature-knowledge a number of investigations are carried out which, in both cases, are much more culture-specific.

In the table we can discern yet another similarity. In Islamic civilisation the steep downturn that marked the end of the Golden Age was not permanent. Later on a certain reversal took place and in three regions nature-knowledge revived: under the Mongols in Persia, the Berbers in Andalusia and the Ottoman Turks in and around Istanbul. All three remained oriented towards their Golden Age, which by then had occurred centuries before. To pursue nature-knowledge meant and continued to mean the production of commentaries on the works of the great figures of the past. From time to time some genuinely brilliant mind appeared, like al-Tusi, who was at least the equal of any scholar of the Golden Age. And yet one consequence of this orientation towards the past was a perennial roundabout – scholars kept going round in circles. When in 1453 the original texts were released from the palaces and monasteries of Byzantium, it brought about a major translation effort in Italy but not in Byzantium/Istanbul itself. Even today nature-knowledge in the Ottoman Empire is focused on what practitioners perceive as their own Golden Age, the one under the Abbasids.

Precisely that circularity, however large the circle might have been, shows a strong similarity to another episode, the upturn of nature-knowledge in medieval Europe. Quite a different cause – that one-sided transplantation – had the same effect of becoming bogged down in one's own intellectual traditions. In both cases, promising initiatives are certainly undertaken, as with the possibility of the Earth's axial rotation. Oresme toyed with the idea, al-Qushji removed every possible consideration that might stand in its way, but that was as far as it went; neither took the next step. In both cases, the transplantation finally ran aground.

Just as in China (although there it was mainly through the absence of any kind of transplantation whatsoever) the development leads in the end to a 'magnificent dead end'.

It is very different with the original flowering in Islamic civilisation and for us in the past century and a half. Here, as there, we are facing an enthusiastic investigation of the wealth of possibilities which the body of Greek knowledge appears to offer. And so Mr Commissioner, I return to my original question. What can the flowering of nature-knowledge in Islamic civilisation, so comparable to our own, teach us in the year of our Lord 1600 about the future of nature-knowledge in the coming years?

In very general terms my answer is as follows: at some point its momentum will slacken, sooner or later there will be a steep downturn. Precisely when? Not much of a sensible answer can be given. Can we expect another wave of destructive invasions, this time in our own region? It is true that, as we all know, the Ottoman Empire has controlled the Balkans for some time and continues to threaten Vienna. All the same, there is nothing to indicate that the great cultural centres in Italy, southern France and southern Germany are at risk. It would make more sense to examine substantive developments within nature-knowledge, and to do that we must draw some careful distinctions.

To start with 'Alexandria'. However enthusiastically enrichment is carried out, it is quite unlikely to go on for very much longer. For in what further direction can it develop? The only natural phenomena that scholars working in the tradition of Ptolemy and Archimedes are concerned with are still equilibrium states, light rays, musical intervals and planetary trajectories, and as before in the usual abstract manner. A new theorem here, a supplementary proof there, but little beyond that seems to be in the offing. I have made discreet enquiries about young, up-and-coming talent and in that respect you should keep an eye on a young schoolmaster in Graz and a highly promising professor of mathematics in Padua. The former, named Johannes Kepler, has just published a book (1596) in which through a weird combination of solid mathematics and imagination run riot he claims to have unravelled nothing less than God's building plan for the universe. The other, a certain Galileo Galilei, left behind an unfinished text when he left the University of Pisa eight years ago. In it he attempted in vain to investigate along the same lines as Archimedes did with the lever the acceleration of falling objects at the beginning of their fall. Precisely because these are two highly ambitious attempts by two extremely gifted young men, their very failure serves to underline my conclusion that the Alexandrian mode has run its course. I have no hesitation in predicting its approaching downturn.

With 'Athens' the situation is slightly, but only slightly, different. Especially the revival of all those dogmatic, speculative intellectual systems! The endless repetition of the same moves on the same chessboard typical of philosophical debate in ancient Greece! It is quite likely that the public, and even the participants themselves, will soon tire of it. True, revisionist attempts to broaden Aristotelian doctrine in the direction of empirical-practical magic (Fernel) or of mathematics (Clavius) have found some followers. Also in terms of content there is potential for further fruitful development. If you have money to distribute, Mr Eurocommissioner, I would suggest you give both these undertakings a modest booster subsidy.

Meanwhile there is a very different development under way from which we may still expect a great deal more. I am thinking of all those practitioners of a 'third' mode of nature-knowledge which is not Greek-intellectual but focuses on accurate observation and practical use. They do not aim to reconstruct an idealised past like their Alexandrian colleagues and most of the Athenians. They are strongly future-oriented. They write their books in the vernacular, not in Latin, and proudly insert the word 'new' into their titles, as for instance the English translation of a Spanish book about medicinal herbs in America entitled Joyfull Newes Out of the Newe Found World. *Here I sense dynamism, a striving for progress, in close contact with the most dynamic elements of society involved in seafaring and overseas trade. There is every sign that this development will continue for quite a while. If we expect nature-knowledge to prolong its flowering anywhere, at least for a while, it is likely to be here.*

To summarise. The preceding survey of the main trends, Mr Commissioner, strongly suggests that you would be well advised to invest most of the funds under your control in this 'third' mode of nature-knowledge. By the way, these also happen to be the men who are most in need of financial support. They have to go out into the world; they need instruments to conduct their observations in the most accurate manner possible. They are not content, as are their colleagues, merely to read old texts and comment on them in the lecture theatre or even, armed with nothing but compass and ruler, sit in their studies doing esoteric mathematics.

The advice is ready, and our reporter may give himself a well-deserved pat on the back for a thorough job well done.

It is rare indeed for a trend-watcher to present such a solidly argued prediction that so quickly proves to be so profoundly wrong.

3   Three revolutionary transformations

Around 1600 the Scientific Revolution broke out. Or, to put it more precisely, a revolution occurred within each of the three modes of nature-knowledge that had flowered during the previous one and a half centuries. But the great paradox of the Scientific Revolution is that the one that underwent the most radical change was the most oriented towards the past. Against all those well-grounded expectations of our trend-watcher, 'Alexandria' turned out not to be nearing the end of its potential for development but really formed the inner core of a fundamental transformation into a largely new mode of nature-knowledge that I shall here call 'Alexandria-plus'. And those responsible for this revolutionary transformation were the same Johannes Kepler and Galileo Galilei who a few years earlier had largely failed in their attempts, undertaken with greater boldness than ever before, to enrich 'Alexandria'. It is through their doing that the time bomb which for half a century had quietly been ticking away in Copernicus' book now suddenly exploded.

KEPLER AND GALILEO: FROM 'ALEXANDRIA' TO 'ALEXANDRIA-PLUS'

Kepler and Galileo never met. One lived in the German-Austrian part of the Habsburg Empire (Graz, Prague, Linz) while the other lived in Italy (Pisa, Padua, Florence). They did exchange a few letters, and Kepler was keen to continue the correspondence, but Galileo held him at arm's length except for the one occasion when he urgently needed Kepler's help. Not only were their characters very different but also their methods and the core problems to which they dedicated their respective careers. What they had in common was intellectual brilliance and a deeply felt need to make mathematics deal with the real world. Kepler derived that desire from Ptolemy, who in his own way had taken some steps in the same direction. For Kepler its urgency was rooted in his conviction that, as Copernicus had argued in Book I half a century earlier, the Earth really is a planet and really does rotate on its axis each day and revolve around the Sun every year. Only, with Copernicus this idea, even if supported in Book I by several arguments, was hard to reconcile with the detailed models that he produced in Books II–VI in the time-honoured 'save the phenomena' style of the *Almagest*. But Kepler, and Galileo, too, perceived, each in his own way, that the kernel of truth hidden in Copernicus' work could be prised loose from the numerous apparent absurdities in which he had wrapped it. The fact that those absurdities were closely bound up with Aristotelian doctrine and really with natural philosophy of any kind did not deter them but rather, particularly in Galileo's case, made the challenge all the more attractive.

Kepler concentrated on the technical-astronomical aspects. In his mind's eye he saw the potential for a model of the universe that was both simple and accurate. The fifty or so auxiliary circles that Copernicus needed to arrange his models so as to make predicted and observed planetary positions match could be reduced to just one orbit each for every planet without weakening the accuracy of the predictions, indeed the very opposite. Galileo, in contrast, was not interested in the countless details that astronomical model building involved. The core of his work lies in a new conception of motion. It enabled him to solve the problem of acceleration in free fall, at first for its own sake, but later also to counter the many common-sense and natural-philosophical objections which for half a century had been levelled at the Copernican system.

In what follows I do not wish to give the impression that they achieved their goals without much intellectual hardship and difficulty. Their initial attempts proved to be dead ends – the very dead ends which I have used to support our trend-watcher's prediction that the enrichment of 'Alexandria' was nearing its end. But what could not possibly have been predicted was that from their initial failures both would draw inspiration to strike out in wholly new directions and ultimately realise their original goals. Exploring those new routes also took them down numerous byways which only in retrospect could and can be recognised as such. Kepler kept a careful record of these 'off-piste' rambles and his most important book is in the form of a report on them. For Galileo they can at best only be reconstructed from his surviving notes. But here is not the place to travel down those side roads. I conceive it as my task to relate the what, the how and also the why of the massive upheaval which the two of them brought about. We do know where they both started: what I have called the third Golden Age of the Alexandrian mode of mathematical nature-knowledge. What we want to know is how far each had moved beyond that starting point by the end of his career. How does what they had achieved by 1630 and 1642, respectively, compare with contemporary approaches and results? To find that out, we need to see only the outcome of their efforts, not so much how they got there. All the same we must view their achievements as products of their own time, not through a hindsight distorted by the later revisions and elucidations of their successors. In considering their respective achievements, then, we shall pretend that the route taken was direct, rather than along the twisting paths which generations of historians of science have since brought to light.

The overhaul that Kepler brought about in the prevailing mode of nature-knowledge is clearly and succinctly defined in the subtitle of what in retrospect proved to be his most important work, published in 1609. The title page reads as follows:

> New Astronomy, based upon causes, or: Celestial Physics, treated by means of commentaries on the motions of the star Mars, from the observations of Tycho Brahe, Gent., by the order and at the expense of Rudolph II, Emperor of the Romans etc., worked out at Prague in a tenacious study lasting many years by His Holy Imperial Majesty's Mathematician Joannes Kepler.

So Kepler's astronomy was a 'celestial physics'. That was not as innocent a concept as it may sound. 'Physics' was just another term for 'natural philosophy', the Athenian approach to natural phenomena, so what business did it have in the realm of mathematical astronomy? 'In trying to prove the Copernican hypothesis from natural-philosophical reasoning, Kepler introduces strange speculations which belong, not in the domain of astronomy, but of natural philosophy', wrote an expert colleague, Peter Crüger, thirteen years later. For Crüger, only the approach practised in the *Almagest* was acceptable and it did not matter if its auxiliary circles could even be imagined as existing in reality. But in Kepler's hands the whole concept of 'physics' begins to shift in the direction of what it has come to mean since. In his *Astronomia nova* Kepler managed to bring about a radical simplification of Copernicus' models in two ways. As the last of Tycho's many assistants, on the famous stargazer's death Kepler obtained his observations and made expert use of them. And, as Crüger observed with such astonishment, his mathematical argument came thoroughly intertwined with 'physical' reasoning. Here the meaning of 'physical' is that he took into consideration the forces at work to keep the Earth and the other planets in their orbits around the Sun.

In the end, Kepler calculated and reasoned his way towards the following solar system: (1) all six planets travel on an elliptical path around the stationary Sun; (2) the slight flattening from circle to ellipse arises from the magnetic action of the Sun, by which it attracts the planet over one section of its orbit and repels it at the other; (3) the Sun, by spinning on its axis, exerts a force on the planets which sweeps them along, as with the spokes of a wheel; (4) a line drawn from the Sun to a planet sweeps out equal areas in equal times; (5) the squares of the orbital periods of any two planets are proportional to the cubes of their average distances from the Sun.

Propositions (1), (4) and (5) are still known as Kepler's three laws. The first two appeared in *Astronomia nova* in 1609, whereas the third occurred to him ten years later. Their full significance would not become apparent until half a century later when Newton discovered the law of universal gravitation. The forces holding Newton's solar system together are not those that Kepler posited in propositions (2) and (3). But valid or not, Kepler's five propositions represent a radical break with the customary practice of astronomy, with the entire Alexandrian tradition of mathematical nature-knowledge and even with the various ways in which Ptolemy had tried to link mathematical science up with the real world – what I have called Ptolemy's bridge-building effort. All auxiliary circles are eliminated (proposition 1), all ambiguities in regard of uniform velocity are eliminated, too, and replaced by one simple area law (proposition 4). Kepler's solar system is unitary and clearly structured, and mathematical expressions have unambiguously defined certain specific properties of the natural world.

And, yet, we should resist the temptation to make Kepler look too modern. The reality for which he formulated his mathematical laws was not merely physical, in the sense of those force actions. It was above all harmonic. Kepler believed that God had created the world according to certain 'world-building ratios' – those which also produce the consonant intervals. In his book on *The Harmony of the World* (1619), which he himself considered to be his masterpiece, he first derived those ratios using advanced geometry. He then showed how God had worked them into music, into the angles which each planet makes with the Earth (Kepler's variant of contemporary astrology) and into the orbital velocities of the planets, which is how he discovered his third law (proposition (5)). His ultimate aim was to prove that, if you thoroughly and consistently calculate your way up to God's design for His creation, the solar system could not possibly have turned out differently from what it is. Mathematics and reality could not be more closely intertwined: the former sets definite limits to the seemingly boundless potential of the latter.

But it all had to be checked. In 1601 when Tycho died and Kepler succeeded him as Emperor Rudolf II's court mathematician he also inherited Tycho's treasure house of unprecedentedly accurate observations. There are still people who believe that Kepler discovered the elliptical shape of the planetary orbits simply by plotting their positions on paper and drawing a line through them. In fact, it did not happen so uncreatively, nor could it have. Kepler needed several years and the most advanced mathematics, which he partially had to develop himself, to hit in the end upon a particular property of planetary orbits which he recognised to be a property of an ellipse. The conceptual problems that he faced along the way were even greater. He first had to free himself from the 2,000-year-old idea that planetary trajectories necessarily have to be analysed in terms of circles, and he had to come to terms with the non-uniformity of their angular velocity (which resulted in his area law). This is not to deny that Tycho's observations were very useful and in particular functioned as a final reality check. For instance, Kepler discarded without hesitation a hypothesis on the orbit of Mars that he had established on the basis of a whole series of observations and of a year's drudgery with the most complex calculations, after checking it against an observation of Tycho's that he had not previously used. What makes this famous event so special is that the observation in question really seemed to confirm his hypothesis quite neatly. After all, it remained within the bounds set by the margin of error valid for Ptolemy and for Copernicus, too, to wit, about 10 minutes of arc (an arc minute is a 21,600th part of a circle). But Tycho had managed to make his observations so accurate as to reduce the margin of error to 2 minutes of arc, so Kepler had to regard the deviation of 8 minutes that he could not fail to notice as sufficient grounds for rejecting his own hypothesis. Or, rather, he did not 'have to' do anything of the sort. The bare fact that he did introduced a quite new element into natureknowledge: *post factum* empirical checks to find out whether a particular intellectual construction can or cannot match up to reality.

Again, we must not portray Kepler as more modern than he was. For him, those elliptical orbits, however proud he was of his discovery, were still only incidental to his central preoccupation with world harmony. In his book of that name so much was at stake for him that he was not always completely ruthless when his derived regularities did not entirely accord with the factual data. By resorting to all kinds of complex *ad hoc* reasoning he managed in the end to prevent God's harmonious creation from collapsing.

Returning from world harmony to the legacy of his years as Tycho's last assistant, Kepler fulfilled a long-standing obligation with the publication in 1627 of the *Rudolphine Tables*. In that book he presented Tycho's collected observations in a systematic fashion that enabled practitioners to prepare tables and almanacs for daily use. And in his elucidations there was enough of his own work to enable a later generation to hive off the three laws of Kepler from a 'physics' that was not really valid and from a vision of world harmony that was much too fanciful.

The physics that did not add up and the fanciful world harmony may well have been part of the reason why Galileo preferred to keep Kepler and his enthusiastic letters at arm's length. In contrast to Kepler's *Harmony of the World* but just like the traditional Alexandrians, Galileo was more interested in individual phenomena than in how they might cohere in some overarching, definitive vision of the natural world. Underlying Kepler's proposition (2) about the force that the Sun must exert to keep the planets in their orbits was the generally accepted notion that to keep something in motion requires the action of a force of some kind. Galileo managed to free himself from that intuitively obvious notion. His supreme achievement was to develop a radically new conception of motion that directly contradicts what our daily experience tells us.

When he left Pisa in 1592 for a better-paid professorship at the university of Padua he had not yet reached that stage. The eighteen years that he spent there before he was appointed court mathematician to the Grand Duke of Tuscany were his most creative period. He published the results only towards the end of his life in two books, one in Italian, the other alternately in Italian and Latin. The first is known as the *Dialogo* ('Dialogue on the Two Chief World Systems, Ptolemaic and Copernican'). It came out in 1632 and gave rise to the famous trial which we shall discuss in the next chapter. Since he could no longer publish anything at all in Italy, his final work, known as the *Discorsi* ('Discourses Concerning Two New Sciences'), appeared in 1638 with Elsevier in Leiden (Holland). Both works cover four 'Days' during which three scholars conduct their learned debates in a Venetian palazzo.

On Day Two of the *Dialogo* Galileo develops the idea that once an object is in uniform motion it tends to retain it. This is quite different from what we observe: if you kick a stone or throw it away it quickly comes to rest. But Galileo now argues that *under ideal circumstances* the stone keeps moving forever. If every impediment were to be removed, you would find that the stone would never come to rest. These impediments rest in such things as the resistance produced by air or by surface friction (for cars or bicycles we speak of air resistance and rolling resistance). Think the obstacles away, and the movement never ends. Imagine that you are rolling an ivory billiard ball over a highly polished marble floor. To be sure, the ball will come to a standstill in the end because even that arrangement does no more than approximate the ideal situation, but it would still take quite a long time.

What all this amounts to is that Galileo distinguishes between three levels of reality: everyday experience, the ideal situation and an experimental level in between.

There is, first of all, reality at the level of day-to-day experience. Every natural philosophy set out to explain everyday reality from first principles. Aristotle was the philosopher who immersed himself most thoroughly in defining the essential nature of motion. What he came up with was no more than our daily experience of moving bodies, but now fitted into the explanatory framework of his philosophy. Motion was for him one of the four manifestations of change. Motion, as change, is the realisation of an end that resides within the moving object itself. Whether or how an object moves is not dependent on the behaviour of any other object. Every movement is on its own; an object cannot perform two different movements at the same time. An object begins to move when the action of an external force makes it do so, and when that force ceases to act the object comes to a standstill (as with a cart when the horse stops pulling).

Galileo does not of course deny that what we observe every day corresponds quite neatly to Aristotle's account of motion. But at the 'ideal' level things are very different. Galileo takes that level of reality from Archimedes and the other Alexandrians. It is the level from which every disruptive circumstance has been abstracted; they have, as it were, been thought away. No air, no surface friction, can stand in the way of ideal motion any more. Looked at in that manner, it is a phenomenon that you do not encounter in the material space of daily experience but only in an ideal geometric space which we can imagine, and even approximate, but never fully realise. If you imagine a moving object in that space, its movement is not directed at any specific end. It moves only in relation to other objects, which are either moving or standing still within that space (like the familiar situation of two trains next to each other in a station: when one of them pulls slowly away there is a moment when you are unsure which train is actually moving, the one where you are seated or the other). Seen in that light there is also no reason why an object should not engage in several different movements at the same time.

In the *Dialogo*, then, Galileo makes a series of closely related statements about motion, which do not at all reflect everyday reality and are valid only in an ideal, mathematical reality. So why should this be of any concern to us?

To that question there are one general and at least three specific answers. The general answer involves the intermediate level. It might perhaps be clearer if we were not to speak of an intermediate level but rather of an escalator, a moving staircase between the everyday and the ideal. That escalator is Experiment. Our ivory billiard ball rolling over a marble floor was such an experiment, as exact an approximation as possible of the ideal level. 'As exact as possible' indeed, for there will always be a difference in practice. Only with our mind's eye, through thought, can we entirely eliminate all disruptive factors. The challenge is to design experimental situations that approximate the ideal situation as closely as possible. And what is that good for? It serves to find out whether the theoretically derived behaviour of moving bodies is actually valid and can teach us something about our own everyday reality.

Galileo saw Copernicus' rotating Earth as a perfect example of how the ideal level of reality can show us what actually happens in the real world. It also provides one of the specific answers to the question just posed. One objection levelled regularly at Copernicus cited the example of a stone being thrown up vertically and landing back where it started. Surely if the Earth were rotating while the stone is in the air we would see it fall back on Earth slightly to the west. This does not happen; therefore the Earth cannot be turning. The argument is absolutely watertight until you realise that it is based on a tacit assumption which is false. Galileo was the first to appreciate this. In the ideal world an object persists in its motion, and can also very well be subject to more than one movement at the same time. And that is actually what happens in the everyday world, too – like the person who throws it up, the stone keeps sharing in the Earth's movement; the movement upwards and then down again is just a temporary addition. It makes no difference whether the Earth is turning or stationary; an object thrown vertically upwards will fall back to the same spot, so at least on this score Copernicus' supposition remains unaffected.

In the *Discorsi*, too, Galileo's new conception of motion is closely tied up with the phenomenon of falling bodies. If you let an object fall, it initially accelerates for a little while until after a few metres its speed no longer changes. So where does the initial acceleration come from? Aristotle, the only classical natural philosopher to take note of this phenomenon, was unable to explain it. Since then, rivers of ink had been spent in an effort to plug this gap in his doctrine. Galileo now begins his argument by proposing to set aside the verbal approach and make the phenomenon fit for calculation. He himself had made a start on it in Padua. He would no longer treat vertical descent as a matter of equilibrium to which Archimedes' law of the lever would apply. He had tried that in Pisa and failed. In the *Discorsi* he now proposed that the speed of a falling body increases uniformly over time. In empty space it would continue to accelerate; it is air resistance that prevents this from happening. From the supposition of uniform acceleration a number of consequences follow for the ratio between distance covered and time meanwhile passed. Some of them could be tested experimentally, even with the extremely limited means available. Galileo made the process controllable by testing it on an inclined plane. He describes in great detail how you should cut a gutter in a piece of hardwood and polish it until it is as smooth as you can make it, so that the test will approximate to the 'ideal' situation as closely as possible. And indeed it has been confirmed by modern reconstruction in accordance with Galileo's own description that the actual and calculated results coincide quite closely.

Galileo went on to use the obtained result for determining the path of a cannonball. What happens when you fire a cannon or throw a stone horizontally? His new conception of motion brought with it the realisation that the stone is now subject to two movements. Horizontally it moves uniformly; vertically it accelerates uniformly. It follows from Apollonius' *Conics* that if you combine the two you get a parabola. Galileo grasped the opportunity to draw a table which, for every angle that a cannon is fired, shows how far away the cannonball would hit the ground. The table shows that it would fire the greatest distance when the barrel is pointed at an angle of 45 degrees to the horizon (i.e. half a right angle). To our earlier question as to what sense it makes to distinguish between the everyday and a mathematically ideal world, this kind of practical application provides a second specific answer: the results can be of great practical use.

There is a third specific answer: the distinction points the way to the artificial creation of entirely new realities. To reach any valid conclusions at the ideal, mathematical level, air is an obstruction that needs to be 'thought away'. It is then no great step to start wondering about ways of actually removing the air or, in other words, of creating a void space. And indeed, as we shall see, within a generation air pumps were being built and put to work which proved capable of producing a virtually air-free space.

As with Kepler, we must be aware of the limitations to which Galileo's radical innovations were subject. One was that he would most often underestimate the discrepancy between the everyday and the ideal. No gunner could have benefited from Galileo's exact tables. Another limitation is that Galileo insisted on motion being retained horizontally, in the literal sense of an object moving parallel to the horizon and therefore in fact in a circle. Not until Newton's *Principia* would it be shown unambiguously that the principle of motion retained, henceforth known as the principle of inertia, is valid only for rectilinear, uniform movement in the absence of any force acting on it from the outside.

But for us it is not so much a question of how far the work of Galileo and Kepler can match up to modern physics and astronomy – from that point of view there are, for all their brilliant insights, still many grounds for dissatisfaction. The question for us is rather how it compares with the work of their immediate predecessors. And then we can see that the distance covered is truly spectacular. Both in depth and in breadth they added a huge 'plus' to 'Alexandria'.

In its depth we can speak of a virtually new knowledge structure. Our distinction between two levels of reality connected by an escalator that can transport you from the one to the other already goes some way towards defining it. The expression that has become customary for conceptualising what all this is about is 'the mathematisation of nature'. At its core is the close connection now created between mathematics and the real world. Galileo was at work all the time to show his readers how the mathematical regularities which he derived theoretically also appear in the everyday reality of a falling object or a discharged bullet – wherever possible he sought to set experimental confirmation in a familiar frame. With Kepler, mathematisation came in the shape of his three laws of planetary motion, his celestial physics and even his use of mathematics as the principal agent to mark out the limits of what is and what is not possible in the real world. Nor did it stop there. Kepler's concern with the reality of natural phenomena enabled him to revolutionise another subject of Alexandrian abstraction: light and vision. Five centuries earlier, Ibn al-Haytham (Alhazen) had based his Ptolemy-inspired conception of light and vision on an assumption to which nobody at the time either in the Islamic world or in Europe had raised any objection. So long as you treat light rays as just geometrical straight lines, the assumption looks quite acceptable. Only when you attach a physical significance to them, as Kepler did, are its weaknesses shown up. And that opened the way for Kepler's radical inversion, by which the lens function of the eye became apparent for the first time.

In terms of breadth it was Galileo who first subjected other natural phenomena to mathematical treatment than the five Alexandrian ones. It had not previously been done with falling bodies, projectiles or motion as such. The same is true of the pendulum for which Galileo discovered a mathematical rule – the time it takes to swing freely to and fro is independent of the width of that swing (its amplitude). The same is true of the strength of materials, the other subject covered in the *Discorsi*. And there are many other phenomena for which (with varying success, to be sure) Galileo managed to derive mathematical rules.

It is true that experimental proof was not always unambiguous or even possible, but that is not the point here. In the early decades of the seventeenth century it was shown for the first time on partly well-trodden, partly novel terrain that the limitations to which 'Alexandria' was subject could be overcome. True, Ptolemy had been aware of the presence of certain limitations, which he sought to meet by way of building bridges of a kind between mathematical science and reality. Ibn al-Haytham and Zarlino had been inspired by his example to undertake something similar in their respective domains of light and vision and of musical harmony. But now, in the hands of Kepler and Galileo, it turned out that mathematics and the real world can be connected much more closely and, moreover, across a much wider range of phenomena than so far. We have every right to speak of a *transformation* in that something that already existed (Alexandria) turned into something largely albeit not entirely different (Alexandria-plus). And we have every right to call that transformation *revolutionary* because it brought with it innovations that were quite as radically unpredictable as any other historical event to which that particular label is routinely affixed.

Just how revolutionary the transformation was can be seen in contemporary reactions. Many of the younger generation, less encumbered by the ballast of outworn habits and assumptions, gave it an enthusiastic reception. But for the great majority of Galileo's contemporaries, who soon became aware that something unusual was afoot, those unusual aspects were rather grounds for rejection. We have already seen the incomprehension of a traditional 'Alexandrian' such as Peter Crüger of something called 'celestial physics'. And, when after Galileo's death his disciples get involved in a lengthy controversy over his rules for falling bodies, time and again it becomes clear that the most innovative 'Athenians' are ready to work with quantities but still have little understanding of the knowledge structure of 'Alexandria-plus'. Time and again the main stumbling block turns out to be the deeply confusing claim that its statements do ultimately refer to the real world but only in a particular, abstract-idealising manner (a source of confusion that science educators have been struggling with ever since).

Another measure of 'revolutionary' lies in the kind of question that the best results of Kepler and Galileo gave rise to in their turn. Before Kepler the question was what circular, uniformly traversed components of a given planet's trajectory were the most appropriate for constructing a mathematical model for that planet. After Kepler, a major issue was how his laws are interconnected. Why ellipses? Why the area law? Why that complicated squared/cubed ratio between orbital periods and distances from the Sun? What do those scattered insights have to do with each other? Before Galileo, the question was why a falling body initially accelerates for a while. After Galileo the question became why its fall accelerates uniformly. In both cases a really unanswerable question is so reshaped and redirected that it becomes viable, even if the answers were not given by Galileo or Kepler themselves – half a century later, Newton managed to distil the answers out of their best findings. Even so, Galileo put clearly into words what was so new about his achievement in this particular regard. At the start of the Third Day of the *Discorsi* in which he deals with the acceleration of falling bodies he announces what the reader may expect:

> The path shall be cleared and made accessible for a very ample, very excellent science, of which these our labours will be the elements, and in the more hidden recesses of which farther-seeing minds than mine shall penetrate.

It is a characteristic mixture of modesty and self-awareness. Galileo sees further than anyone previously, but he generously acknowledges that he cannot see all the way to the end. That end lies beyond the horizon in an unknown future. Nevertheless, that deliberate drive to innovate has now begun to affect the 'Alexandrian' approach. Previously it had only been practitioners of the descriptive, practice-oriented mode of nature-knowledge who injected a self-conscious 'new' into the titles of their books (for instance, *Joyfull Newes ...*). Now Galileo in the very title of his *Discorsi* refers to '*new* sciences' while Kepler entitles his book on celestial physics '*New* Astronomy'. Now what made such revolutionary innovation possible, and how is it that, against the well-reasoned expectations of our trend-watcher, it happened precisely in Europe?

We already know the answer to the question of what made it possible. The transformation was present as a latent possibility in the Greek

and in particular in the Alexandrian legacy. Not only had the characteristic approach of abstract-mathematical idealisation already made
its entrance, but Ptolemy had perceived that the level of abstraction
was exceedingly high and had made efforts to tie the mathematical
mode of nature-knowledge more closely to the real world. The greatness of Kepler and Galileo lies in their actually achieving that goal,
not in any claim to be the first to attempt to reach it. And when we set
out to explain how they managed actually to attain that end, the
prime circumstance to point at is indeed that this feat resided as a
potential development in the specific cultural transplantation which
had taken place in the previous century and a half.

Does this imply that each of the three transplantations that we
have considered might have had a comparable outcome, the emergence of 'Alexandria-plus'? To a certain extent this is indeed so.

In the first place, at each successive transplantation of the
Alexandrian legacy the chance increased that its development potential would be realised. That applies in particular to specific aspects of
enrichment which having been acquired on one occasion could affect
processes of revival and enrichment at the next. And although the
mathematics that Kepler and Galileo employed for their great breakthroughs was primarily Greek geometry, the impact of 'Arabic'
numerals and the availability of some algebra surely had an effect on
Kepler's work in particular. Also the Tusi couple, at least if Copernicus did not re-invent it independently, may have indirectly contributed to the breakthrough.

More important, however, is the strong similarity between the
revival of Greek nature-knowledge in Islamic civilisation and in
Renaissance Europe that we noted in the previous chapter. The like-
ness applies particularly to the first upswing in Baghdad, rather than
to the later revivals in Persia, Andalusia and the Ottoman Empire.
Those late 'Islamic' revivals instead display similarities with the
second cultural transplantation to medieval Europe: whether it was
an orientation towards their own Golden Age, in the one case, or
towards exclusively Aristotelian doctrine in the other, the pursuit of

nature-knowledge kept going round in (admittedly large) circles. Things were far more open at the time of the original revival of the near-complete corpus of Greek nature-knowledge. In the process of rediscovery, translation, appropriation and enrichment that then ensued, we sense drive, momentum, enthusiasm, a genuine thrill of discovery. An inner dynamic was unfolded in which whatever potential for development that might be present at least had a chance to come into the open. Or to put it more precisely, the kind of breakthrough that Kepler and Galileo brought about as the culmination of one and a half centuries of Alexandrian enrichment was not in principle beyond the powers of the generation that followed such high-achieving Alexandrians as al-Biruni or Ibn al-Haytham. Imagine that in the critical period following their deaths in around 1050 (a decisive moment in retrospect) any further development had *not* been stifled by the wave of invasions that began by then to upset the Islamic world. In such a case nothing would have stood in the way of the appearance of a Galileo-like individual in Islamic civilisation.

This looks like a fairly wild assertion. But empirical backing for it is provided by the high degree of similarity between these two Golden Ages of Alexandrian nature-knowledge. Why should the immediate follow-up in the one case have of necessity been beyond the reach of the other? Let us, by way of further support, take a closer look at Galileo's intellectual development (with Kepler things were more complex but not essentially different). Galileo began his career as a typical and even a conspicuously purist Alexandrian. His first piece of research was the attempted reconstruction of a problem that had exercised Archimedes. He was, further, more familiar than most Alexandrians with the teachings of Aristotle from whom he took his growing interest in falling bodies and especially in the problem of acceleration in free fall. His first attempt to pin it down mathematically got nowhere – Archimedes' law of the lever can be of no help in that regard. Even so a few isolated sentences in his unfinished treatise on the subject contain the seeds of an idea that later on in Padua would lead him to his conception of motion retained. The step that he took from a pure Archimedean approach to one belonging to what I have called 'Alexandria-plus' was surely neither small nor gradual. Yet it is possible to break it down into a number of smaller steps, and Galileo's handwritten notes make it clear that he did indeed take them in succession. In a different situation and in other circumstances those intermediate steps might well have been taken in a different way or divided between different scholars working together or in succession. By 'Galileo-like individual' I therefore emphatically do not mean that in the absence of large-scale invasion things would have happened in the Islamic world exactly as four and a half centuries later in Europe. But I do mean that the final result, the realist-mathematical mode of nature-knowledge of Kepler and Galileo, might under non-disruptive circumstances have been reached there as well, even if no doubt by a more or less different route.

The core point of this whole argument is that there was no inherent necessity for the revolutionary transformation from 'Alexandria' to 'Alexandria-plus' to take place in Europe. It could just as well *not* have happened. It could also have happened somewhere else, around 1050 in Islamic civilisation as the culmination of its Golden Age of nature-knowledge.

We may still point at a number of circumstances which, while not *determining* that such a breakthrough would in fact happen in late Renaissance Europe, nevertheless increased the chance of its happening.

Kepler and Galileo were exceptional individuals of such tremendous gifts that it is at least conceivable that without them the revolutionary transformation would not have happened at all. However, genius on its own is never a sufficient explanation. Talent for mathematics and for the pursuit of nature-knowledge is of all times and places; the question is whether it will be granted a chance to express itself. If it is not at least to some extent anchored in society it cannot find expression. While nature-knowledge was flowering in Islamic civilisation opportunities for radical change were certainly present, yet in Renaissance Europe they were significantly greater. That was due primarily to the presence of universities that provided the European elite with an education in which nature-knowledge formed a significant part. It is true that the Aristotelianism that was taught there pointed in a quite different direction. Nonetheless, many a student received enough elementary mathematical knowledge to awaken their interest, and indeed if sufficiently gifted to take it further. As noted, Kepler was Tycho's last assistant and at least thirty others preceded him. However much training each received from Tycho, on arrival at Uraniborg they all had at their disposal at least the basic mathematical knowledge required for the job. Such numbers were not attained in Islamic civilisation (the ratio of mathematical practitioners in Islamic civilisation as compared to European was about 1:4).

In contrast, another difference between the two civilisations, the printing press, hardly enhanced the chances of a drastic transformation. In Islamic civilisation texts circulated briskly and at affordable prices. Knowledge acquired from manuscripts was not necessarily inferior in quantity or quality to that acquired from printed books. The erudition of scholars in Islam was often quite as formidable as could be found in medieval or Renaissance Europe, and in a general sense their learning would compare well with that of Kepler and Galileo. As we shall see later, the printing press *did* have a decisive impact on *other*, later revolutionary transformations among those six that, together, constitute the Scientific Revolution. But the specific transformation from Alexandria into Alexandria-plus could have occurred in a script culture almost as well.

There remains the question of how it came to that 'plus'. 'Plus', to recall, stands for the novel quality of realism that the abstract-mathematical mode of nature-knowledge acquired in Kepler's and Galileo's hands. Three circumstances that we encounter in Europe but not in the Islamic world may have contributed to it.

The first is the realism that Copernicus displayed in Book I of *De revolutionibus* albeit hardly in the other five Books. In the context of the whole work it was therefore an incoherent realism that evoked among his contemporaries and even in the next generation mostly rejection or, at best, partial and inconsequential acceptance. Kepler and Galileo were the only ones to accept the realism of Book I completely and with a clear awareness that the context in which Copernicus had embedded it was in urgent need of revision. Had it not been for these two men, the chance for radical transformation latent in Copernicus' odd claim to realism for his supposition of a moving Earth might easily have been missed (as in fact it was by their nine fellow Copernicans).

Another possible source of realism lay in what we have called the third mode of European nature-knowledge, which featured investigators dedicated to accurate observation and practical improvement. In contrast to the 'Alexandrians', everything hinged on the realist quality of their investigations. I have argued earlier that, as a rule, practitioners of the three modes of nature-knowledge worked independently of each other. Mutual isolation set the tone, even at times within a single brain (as for instance that of Castro). But Europe offered a compact environment which could be travelled across relatively easily, so more than elsewhere scholars of different stripes were to some degree aware of each other's work. In astronomy this closeness was felt most strongly, and for Kepler the possession of Tycho's recorded observations naturally made a huge difference. But for Galileo too there was a link between these widely disparate modes of nature-knowledge. He could not have taken from elsewhere the kind of experimental confirmation for his rules of falling bodies, since nobody had ever done such a thing before. But coherent ranges of experiments of a different type had on occasion been conducted previously. They were not, as Galileo's were, directed at confirming a regularity that had been derived beforehand, but at uncovering as yet unknown aspects of nature. We encountered such sustained ranges of exploratory experimentation with Leonardo (friction) and Castro (measurement). Before 1600 that is as far as it went – with one exception and that exception lay literally within Galileo's grasp. His father Vincenzo Galilei was more than just an original composer who contributed much to the rise of the baroque style in music. He also did research into the role of the consonances in music making and set up a series of experiments to investigate what effect varying factors such as the thickness or the material of a vibrating string have on the intervals produced thereby. Vincenzo's eldest son must have watched these experiments from close by and derived inspiration from them when, at a much later stage, he sought to place the phenomenon of falling bodies on a new, mathematical footing. In both the *Dialogo* and the *Discorsi* he even tried his hand at experiments of the exploratory kind, albeit more in a style of clever individual invention than of systematically undertaken progressive series as with his father.

Finally, it may seem as if the mixed mathematics which the Jesuits so assiduously practised must also have been a source of realist inspiration. After all, there is a superficial resemblance to what Kepler and Galileo came up with: calculating with real phenomena. And it is possible indeed that such activities encouraged both men to continue along the path which they had taken. However, the difference in knowledge structure is far too great for any genuine affinity. The true greatness of Kepler's and Galileo's intellectual adventure is that they set out on a bold quest with an unknown and unpredictable outcome. With 'mixed mathematics', by contrast, the outcome was already known since it lay firmly entrenched in the first principles of Aristotle's intellectual schema. For practitioners of mixed mathematics the only open-ended question was how far in any given case it would prove possible to expand it quantitatively.

An Aristotelian, whether quantitatively inclined or not, already knew beforehand how the world fits together; only subordinate portions thereof were still open to further investigation. For Kepler and even more so for Galileo, in radical contrast, the whole world lay before them virtually undiscovered. I said this before about the ancient Greeks and Chinese, and now, behold, it seemed to be happening all over again. Both men expressed a profound conviction that the composition of the world is, at bottom, mathematical. Galileo called mathematics the language in which the Book of Nature was written. Kepler spoke of geometry as eternal and as providing God with the very model for His Creation. In their vision, the investigation of nature was still in its infancy, now that in mathematics they had hit upon the pre-eminent tool for re-creating the world.

BEECKMAN AND DESCARTES: FROM 'ATHENS' TO 'ATHENS-PLUS'

Some ten to twenty years later, elsewhere in Europe an attempt was made to re-create the world along quite a different route, not mathematical but philosophical. This time it is ancient atomist doctrine which is radically transformed, and there is every reason to call the event revolutionary as well, even if to a lesser degree. Unlike the transformation of Alexandria into Alexandria-plus, 'Athens-plus' retains the original knowledge structure intact. The philosophical first principles certainly undergo a substantial change, but they are still taken to be all-embracing, and indisputable certainty is still claimed for the foundation of all nature-knowledge. Also, here again according to the pioneers each individual natural phenomenon lends itself to explanation from those first principles and no other.

The great pioneer of this revolutionary transformation was René Descartes, although he was not the first. The first was his older friend Isaac Beeckman. The fact that he has remained unknown is largely due to the fact that this native of Zeeland, builder of water conduits, candle-maker and school teacher, never wrote down his conception of the natural world in one coherent argument. Nevertheless, from the diary he started to keep as a theology student in Leiden it is possible to reconstruct a reasonably consistent natural philosophy which certainly had an impact, especially on Descartes. They met each other in 1618 in Breda when René was garrisoned there as a young soldier in Prince Maurice of Orange's army and Isaac, who was ten years older, had come over from Middelburg to help his uncle and 'for courtship as well' (indeed he would later marry the lady in question). As a thinker, Descartes was definitely sharper, more organised and philosophically much more thoroughly schooled than

Beeckman. The former was the genius, the latter just highly gifted, and it may well be that even without Beeckman Descartes would have designed his natural philosophy just the same. Not that Descartes himself dared to trust posterity to reach this judgement. When he returned to the Netherlands in 1628 in order to commit his now fully ripened philosophy to paper, he found that his older friend had like-wise begun to systematise in book form thoughts so far entrusted to his diary only. In two letters which are as venomous as they are mendacious, Descartes managed to exploit his over-modest friend's lack of self-confidence to the point where the latter gave up forever on his one and only attempt to write down his natural philosophy. After his death the diary went missing and was found only at the beginning of the twentieth century; it was finally published fifty years after that.

It is easier to see what was actually transformed from Beeckman's diary than from the published work of Descartes, who used to present himself as the man who had thought everything through from scratch, owing nothing to any predecessor at all – not a re-creation, then, but single-handed creation.

Both Beeckman and Descartes had a vision of the world as consisting of nothing but invisibly small, impenetrably hard, poly-morphous bits of matter which come together temporarily in clusters before continuing as before in ceaseless movement. So far, nothing new; that is precisely what the ancient atomists had proposed. Until then, atomism had not enjoyed a particularly prominent position among the quartet of Athenian natural philosophies. In the later Roman Republic and the early empire, the Stoa had been predomin-ant; in the late empire it was a variety of Platonism. In Islamic civilisation, Aristotelianism rose to prominence alongside it, and achieved the dominant position in al-Andalus as well as in medieval Europe. It retained that position in Renaissance Europe even though Aristotelians now had to put up with the re-emergence of the old competitors. Among these was, of course, atomist doctrine, which in Islamic civilisation had played a subordinate role in speculative theology, and in Europe had reappeared briefly in the time of

Charlemagne. Twenty years after the fall of Byzantium, the authoritative text in which ancient atomist doctrine is extensively recorded had already appeared, Lucretius' poetic work *De rerum natura* ('On the Nature of Things'). But only in the late sixteenth and early seventeenth century did the work start to have an impact, when a few individual scholars began to nudge Aristotle's account of matter in the direction of atomism. Beeckman was one of them, though quite early on, in about 1610, he introduced a twist of his own. That twist was about how atoms move. In the atomist doctrine of ancient Greece attention was focused on the particles themselves, their size, their hardness and in particular their shape. Thus Lucretius explained the bitter taste of certain stuffs by the sharp, pointed shape of the particles out of which they are made. They quite literally prick our tongue to pieces. But how atoms actually move was of little concern.

To Beeckman it was. Quite independently of Galileo he developed a remarkably similar conception of motion. 'What has once started to move moves forever unless it is prevented', he noted repeatedly in his notebook from 1613 onwards. For Galileo this rule applied to uniform movement parallel to the horizon which is therefore circular. In Beeckman's view an object would keep moving uniformly in a straight line, or in a circle, until something prevented it. Against the background of this conception of motion he now investigated a large variety of natural phenomena with the aim of explaining them from the interplay between his first principles.

As a rule, these explanations were much more specific and more detailed than in ancient atomism. There, for instance, sound had been conceived of as a stream of particles which, after being emitted by the voice or other agent that produces it, reach our organ of hearing and there make themselves felt somehow as sound. In Beeckman's view, sound arises because a vibrating string or other object cuts the air up into small particles, sending them in all directions. The faster the vibrations break up the air into particles, the smaller these become, the faster they fly out, and the more fiercely and quickly they ruffle our eardrum. Beeckman continues up to the level of the nerves and the brain this explanation of how faster vibrations divide the air up more finely, thus producing higher notes. Similarly, a larger quantity of particles sent out by a vibrating string produce the sensation of a correspondingly louder sound. So all the way there is a one-to-one explanation: any given phenomenon that we observe in the macro-world corresponds neatly to a specific corpuscular mechanism in the micro-world.

A phenomenon that is still called 'sympathetic resonance' came in for explanation along similar lines. When you pluck a string, it may happen that a nearby string begins to sound as well, to all appearances quite spontaneously. This, together with the mysterious attraction of iron by a magnet, was the phenomenon most commonly used to illustrate the secret correspondences and the occult forces that were so prominent in the magical view of the world. But Beeckman perceived that there is nothing occult about the phenomenon and that 'sympathy' has nothing to do with it. What actually happens, he argued, is that the vibrations of the plucked string make the surrounding air rarefy and condense in quick alternation, thus causing a nearby string to vibrate in its turn.

To the modern physicist the last explanation is fairly close, whereas the idea of sound consisting of particles of air sliced up by a vibrating string is sheer nonsense. To the historian, the main issue lies elsewhere. This new focus on the specific movements of particles of matter enabled natural philosophers to provide explanations quite a bit closer to what we recognise as 'physics' than was the case with ancient atomist doctrine. This is so because the mechanisms that go into the explanation are far more specific. Whether this or that explanation is really valid is not the important point. Indeed, within the 'Athenian' knowledge structure hardly any means were available for distinguishing between valid and not valid. The particles and their movements could not be seen with the naked eye, but only imagined. They follow on from first principles, but how they might work out at the level of individual phenomena is still open to wide differences of opinion.

Descartes, for instance, thought quite differently of particles in space than Beeckman did. Beeckman followed the ancient atomists in accepting that atoms move through void space, but Descartes did not accept the existence of a vacuum. For him, matter and space are identical. Furthermore, the total amount of motion in the universe is constant – what is added here is taken away there. So a particle can move only by changing places. One particle pushes another particle away, which in turn pushes another out of the way, and in this manner particles group themselves into smaller or larger whirlpools or vortices of matter. In his *Principia philosophiae* (1644) where all this is expounded, these vortices play a large part in the explanations of phenomena which fill the book. He sets out the basic principles of his natural philosophy, from which all his explanations are derived, much more stringently and systematically than Beeckman had done. The conservation of the total amount of motion, the rule that an object will maintain its uniform motion in a straight line, and the relativity of motion are assertions which for Descartes have the status of natural laws. That was then a new concept, which Descartes used for those natural regularities which operate uniformly and everywhere in the universe. The idea of natural law is among his most important contributions to the Scientific Revolution.

It is this new conception of motion that gives to classical atomism a new vitality and at least some resemblance to what we now call physics. Therein lies much of the 'plus' in the revolutionary transition from 'Athens' to 'Athens-plus'. The question is: how could that 'plus' come about?

Just like 'Alexandria-plus', the core of the answer is that what had long been latent in the 'Athenian' heritage now actually emerged. But the *chance* of its happening was somewhat different from the case of Alexandria. The first serious opportunity to transform ancient atomist doctrine occurred in Renaissance Europe. The accidents of text survival are decisive here. Due to these accidents, what little was known about the doctrine in Islamic civilisation stemmed solely from Aristotle's arguments against it. And the speculations of certain Islamic theologians about the atomic structure of matter remained wholly limited to first principles. To link these up with natural phenomena that might have lent themselves to an explanation in atomist terms was far from their purview.

Three circumstances increased the odds in favour of Renaissance Europe. Europe had the core text at its disposal, Lucretius' poem 'On the Nature of Things'. Europe also had universities where for generations the emergent elite had followed a solid course on natural philosophy. True, as a rule it was centred on the teachings of Aristotle. Yet it provided some training in philosophical thinking in any case and, the more individuals made their acquaintance with it, the greater the chance that critical spirits would take it further. This chance was heightened in particular when Aristotle lost his monopoly and his old rivals in natural philosophy reappeared on stage. The intellectual development of Beeckman and Descartes in Leiden and at the Jesuit College at La Flèche, respectively, displays the process quite clearly. Even so, adopting a critical stance toward the dominant form of natural philosophy is still not the same thing as opting for another one, let alone the radical transformation thereof. At most, we can say in this regard that atomist doctrine was more open to transformation than its rivals in so far as it was less spiritualised than Platonism and in its basic principles somewhat more flexible than either the Stoa or the teachings of Aristotle.

Even so we still have to explain where the new conception of motion that constituted the 'plus' actually came from. Of course it was not entirely new, in so far as twenty years before Beeckman first and then Descartes started working on it, it had already been excogitated by Galileo albeit, of course, in an 'Alexandrian' framework. How could those two 'Athenians' have come to the same conclusion?

In the case of Descartes, the question is easier to answer than for Beeckman. Descartes encountered the idea in Breda in 1618 in conversations with his new friend. How Beeckman first hit on the idea remains rather an enigma. The assertion 'what has once started to move moves forever unless it is prevented' appears very early on in Beeckman's diary without further clarification or context. The possibility that he took the idea of motion retained from Galileo can be ruled out – by 1613 Galileo had not yet made it public, either by word of mouth or in published writing. Furthermore, Beeckman always noted scrupulously in his journal where and from whom he had obtained an idea. So all we can say in this regard is that in the span of about two decades two very contrasting thinkers who operated in two widely different modes of nature-knowledge arrived along different routes at the same fundamental insight. Did perhaps late Renaissance Europe provide a particularly fertile soil for a find like this? The question will admit of a better answer once we have made ourselves acquainted with yet another – a third – revolutionary transformation which took place at the same time.

BACON, GILBERT, HARVEY, VAN HELMONT: FROM OBSERVATION TO EXPERIMENT

Halfway through the fifteenth century, the fall of Constantinople had brought about a fresh transplantation of the two Greek modes of nature-knowledge. In its slipstream a third mode of nature-knowledge made its appearance, which was characterised instead by painstakingly accurate observation and a strong orientation towards practical usefulness. This mode of nature-knowledge also went through a revolutionary transformation around 1600. It was brought about by two Englishmen and a Southern Netherlander. It was supplied with a visionary programme and detailed methodology by yet another Englishman, no less a person than Francis Bacon Lord Verulam, Chancellor to King James I, until he fell from grace.

Bacon had little time for the two Greek modes of nature-knowledge. He ignored 'Alexandria' and was unaware of the revolutionary transformation it was going through. He railed against 'Athens' in many of his writings because of its intellectualist top-down approach to natural phenomena. In his view, the pursuit of nature-knowledge was a bottom-up process, not based on predetermined first principles but on open-minded observation along the lines laid down by practitioners of the third mode of nature-knowledge. Except that as such it lacked coherence and objective: like ants, observers kept piling one small piece of information upon another without ever turning them into something truly solid. The collection of knowledge should be undertaken systematically, following the example of the bees. Research should be organised collectively; the material should be worked on methodically. The honey thus obtained was, after all, what all the effort was about. Knowledge of nature, Bacon never tired of arguing, was both the means of and the condition for improving human destiny. In so far as man can undo Adam's Original Sin here on Earth, it will have to be done through nature-knowledge. Its ultimate goal is 'the effecting of all things possible', as he expressed it in one of those immortal one-liners in which he excelled. But the opposite was also true; nature could be mastered only on the basis of insight into its workings: 'Nature to be commanded must be obeyed.'

In all this, teamwork was key. Craftsmen need to work together all the time, and it was primarily from their example that Bacon insisted that it was teams rather than individual investigators who could make the most of his large-scale programme. In his utopian *New Atlantis*, he sketched a society whose central institution is the 'House of Solomon' where knowledge is systematically collected and analysed. There, as many as eighteen types of official are assumed to work closely together in a strictly ordered hierarchy. For instance, the House sends out 'merchants of light' to gather knowledge which other peoples have acquired (really an early form of international research espionage).

The methodological rules which Bacon devised amounted to drawing up exhaustive lists of circumstances in which a particular phenomenon, heat for example, might occur: in sunlight or in various chemical reactions, and so on. On another list you note down under what otherwise similar circumstances it does not occur, as for instance moonlight, which is not associated with heat. Through a systematic comparison of the two lists you may arrive at an initial generalisation. Heat, for instance, always seems to appear together with movement. With such first-order generalisations you can then move up a level, draw up new lists and repeat the procedure until you have finally risen to the most general rules which it is possible to detect in nature.

For those unfortunate enough to take it literally, this methodology soon proved to be unworkable. But that was not true of its most innovative feature. In order to achieve this systematic correlation of phenomena, Bacon proposed to put nature to the trial very nearly in the juridical sense of the term. If nature proves unwilling to give up its secrets to spontaneous observation, you must devise an experimental setup which will compel nature to display its properties with artificial means. It is a process of exploratory, fact-finding experimentation that is fundamentally different from the confirmation-directed experiments that Galileo employed at the time to forge a union between mathematics and the natural world. Again the goal is practical improvement. One example is Bacon's advocacy of experiments designed to bring about the artificial 'majoration' of sound, by means of devices to make sounds louder or carry farther. Speaking trumpets, 'ear spectacles', echoes and whispering galleries all found a place in Bacon's programme for a natural history of sound and how to make proper use of it.

Bacon was a good example of someone who preaches what he himself fails to practise. He filled entire books describing how to set up experiments but he overcame only rarely his reluctance to soil his hands by carrying them out himself. With but little exaggeration William Harvey, physician to Charles I, mocked Bacon for approaching nature like a genuine Lord Chancellor.

Harvey himself was a skilful experimenter. Around the turn of the century he went to study medicine in Padua, where he learnt of the latest discoveries that Vesalius and his successors had made through carefully accurate observation. Back in England he elaborated on them in so far as his busy practice would allow. He saw more acutely than any of his teachers had done that the recent discoveries about the heart and blood vessels were hard to reconcile with the accepted schema of the workings of the various parts of the body. The Roman physician Galen had conceived of blood as a life-giving fluid which is made from food in the liver and then transformed by the heart and the brain into, respectively, vital spirits and animal spirits. The remaining blood flows out through the arteries to the various parts of the body, which extract food from it; the remains perish. All of this presupposed a passage for the blood from the right to the left ventricle of the heart. Vesalius found out that no such passage exists. Luckily, Galen's doctrine of bodily functions, which otherwise managed quite well to arrange the known facts into one coherent order, was saved by the discovery of pulmonary circulation, from the right to the left ventricle via the lungs.

But there was more to worry about. Harvey concentrated in particular on the heart and discovered through vivisection, mainly on dogs, that the prime movement of the heart is not its expansion, as in Galen's system, but its contraction. Subsequently, he calculated in a deliberately conservative estimation how much blood the heart forces into the arteries with each contraction. It turned out that in less than half an hour it would be more than the amount of blood contained in the entire body. Where else can all that blood go to than back through the veins to the heart? In short, then, the blood circulates in a closed system and the heart works as a pump. But the heart is much more than a pump. According to Harvey it is the foundation of life, it purifies the blood and thereby feeds and maintains the whole body in a continuous cycle that in miniature reflects how the Sun orbits the Earth. In this, and in his later work on reproduction, Harvey's world-view is of the 'vitalist' type. In that view no distinction is made between living and non-living nature, and natural phenomena are explained in the end as the workings of some creative, life-giving primal force.

In 1628 Harvey published his discovery in a book entitled *On the Motion of the Heart and Blood*. It contains a characteristic combination of three aspects in interaction: exploratory experimental research, suppositions about the way in which phenomena might be connected, and an underlying conception of the world as a living, animate organism. The same combination can be found in the work of the two other pioneers of exploratory, fact-finding experiment, William Gilbert and Jean Baptiste van Helmont.

Gilbert was physician to James I's predecessor, Queen Elizabeth. In 1600 he published a book entitled *De magnete* ('On the Magnet'). Using Pierre de Maricourt's pioneering work of almost four centuries earlier, he developed it at length and in a systematically experimental way. Thus he found that the distance across which a magnet can make a piece of iron change direction is greater than the distance over which it can still be attracted. This led him to speak of a magnet's 'sphere of activity', a fairly loose idea that displays some family resemblance to the modern field concept. Some experiments, of which he learned from others, involved the compass needle and its several deviations from the geographic North Pole. One of his chief aims was to demonstrate the essential difference between two phenomena that the Greeks had found equally mystifying and had often lumped together: the attraction of iron by lodestone, and the attraction of snippets of paper attained by rubbing amber (in Greek, *elektron*). This way he found that magnetism involves only iron, but that 'electric' attraction may also come from rubbing glass or sulphur.

The full title gives a more ample idea of what Gilbert hoped to achieve: *On the Magnet and Magnetic Bodies, and on the Great Magnet the Earth; a new Physiology, demonstrated by many arguments & experiments*. Gilbert was the first to argue that the Earth is a giant lodestone, contaminated by a few impurities at its crust only. Nature is animate and wholly filled with life; the Earth's magnetism is the form in which the world soul is expressed.

The same idea that the world is animate was central to the chemical doctrine of Paracelsus, albeit in a different way. That doctrine, too, was transformed experimentally during the same period. 'Transformed' indeed, in that Jean Baptiste van Helmont modified certain portions of the doctrine without altering its core tenet, that human beings (the microcosm) and the universe (the macrocosm) alike operate by means of the three 'sophic principles' of sulphur, mercury and salt. Particularly innovative was van Helmont's unstinting effort to give Paracelsus' conclusions a quantitative and experimental twist. For instance, he investigated with great care how acids and alkalis neutralise each other's action. He did not confine himself to chemical reactions in glass test tubes. He caught a drop of fluid from a sparrow's stomach on his own tongue which helped him to find that the same process of neutralisation plays a role in digestion, too. Meanwhile van Helmont was also very much a vitalist, persuaded that natural events are ultimately brought about by certain 'life-giving seeds'.

Whatever the differences in how these four pioneers of exploratory experimental research operated, they all shared a view of the world as animate, a strong orientation towards practicality and craftsmanship, a readiness to challenge nature to produce phenomena which do not appear spontaneously and (except for Bacon) an ability to perceive interconnections between those phenomena. On occasion they employed relatively uncomplicated instruments. Gilbert, for instance, made a magnetic needle swing freely in a vertical plane and thus discovered its inclination to vary with geographical latitude.

None of this was without precedent. We have come across coherent ranges of experiments of the fact-finding type in three cases, those of Leonardo da Vinci, João de Castro and Vincenzo Galilei. But now, around about 1600, they are set up more widely and more systematically, within a broad, often magical and always vitalist world-view which, as such, remained largely intact. Of the three revolutionary transformations that we are exploring here, this is – relatively – the least radical one, in that it remains more closely than the others in line with what directly preceded it. The question of why this revolution occurred can therefore be answered fairly simply. More directly than in the cases of 'Alexandria-plus' and 'Athens-plus' this particular revolution was present as a latent development potential in its immediate predecessor, Europe's 'third' mode of nature-knowledge. In a gradual process around 1600, practically oriented, accurate observation is condensed into fact-finding experiment, where artifices are increasingly employed to force the manifestation of natural phenomena which would not have appeared of their own accord. In the present case, then, our trend-watcher was least mistaken. In 1600 we heard him (for it will certainly be a him in this time period) derive from current trends that this dynamic, future-oriented mode of nature-knowledge might well continue to flourish for quite a while yet. It is only the revolutionary turn towards a much more systematically undertaken experimental approach that he was, of course, unable to foresee.

He was far more thoroughly mistaken in the case of the other two revolutionary transformations: the new *realism* of the mathematical mode of nature-knowledge, and the kind of explanatory mechanisms that came within the reach of natural philosophy by focusing on the *motion* of particles of matter. Above all, our trend-watcher could not foresee that there would be any revolutionary transformations at all in the early seventeenth century and certainly not three at virtually the same time. We have found explanations for each of them, which for the largest part turned out to be different for each individually. But how can it be that the three revolutions broke out at almost the same time? That is the question that still needs to be answered and that we shall consider now.

WHY EUROPE?

The rise of modern science in the seventeenth century and the modern technologies made possible thereby in the nineteenth were instrumental in securing for Europe two acquisitions without any precedent: wealth and luxury for more than just a small elite, and domination over the rest of the world. Since then wealth and luxury have spread to a growing number of the world's inhabitants and little is now left of that early domination. But much of the image that Europe formed of itself in the nineteenth century has remained part of our collective mental framework. At the core of that image lies a belief that Europe was superior to all other civilisations, for how else could Europe have attained its wealth and dominance? But wherein Europe's superiority, or conversely the inferiority of other civilisations, lay precisely has over time given rise to much debate and widely ranging points of view. The decisive difference has been sought in the 'white race', or in Europe's uniquely early capitalism, or in Western pluralism as against Oriental despotism, or in Europe's individualist cult of the personality versus more collectivist and conformist orientations elsewhere, or in the dynamism of a new, rising civilisation compared with the static character of all those older, sleepy ones, or in a combination of some or all of such points of view. And the seeds of all of this have subsequently been sought in European history, going as far back as the Renaissance or even the early Middle Ages.

In recent years, a branch of history known as global history has arisen, whose practitioners (economic historians mostly) are busily challenging the received image at its very core. The explanation by racial difference has long ago been consigned to a well-deserved dustbin. It is also becoming more and more clear that before the early nineteenth century there were no structural differences between the great civilisations of China, Japan, India, the Ottoman Empire and Europe profound enough to lead ineluctably to the wide divergence that then came about. Naturally there were considerable differences in the Old World, but these have existed between every pair of civilisations and cannot be simplified to some polar opposition between 'the West' and some undifferentiated 'Rest'. In Europe, the state was not so weak and elsewhere not so strong as to explain so radically different an outcome. Economic historians, in particular, have meanwhile demonstrated as well that little remains of the allegedly vastly greater poverty, let alone the supposed indolence, of civilisations in the East prior to the great divergence of the early nineteenth century.

Much in these conclusions seems to me enlightening and often also convincing. Above all, I concur with the essence, the effort to release history from the hold of that hoary picture. I have therefore looked out throughout this book for whatever was of by and large equal value in the pursuit of nature-knowledge in China, the Islamic world and Europe. In my various explanations of the three revolutionary transformations that together make up the first stage of the Scientific Revolution I have also deliberately kept my distance from references to European superiority and possible grounds for it.

Or, at least, I have done so as far as I could. But here, at the point which we have now reached, is as far as it goes. After all, not *everything* in the older picture is untenable or irrelevant. We have traced all three transformations separately back to their respective origins and we are left with the question of why they occurred not only within the same civilisation but also at the same time. And that forces us to some further reflection on Europe's peculiar nature during that period. This requires a prior statement about the making of value judgements in connection with the method of cross-cultural comparison that I have followed throughout this book.

I emphatically do not take such distinctively European traits as I am about to invoke and whose effects I am about to analyse to be of a higher (or, for that matter, of a lower) order than their counterparts in other advanced civilisations. None of this, to be sure, is equivalent to a renunciation of all value judgements. I do not subscribe to a posture of wholesale cultural relativism; I rather think that just about every advanced civilisation has been superior in some specific regards, inferior in certain specific others. Europe, although not superior across the board, has definitely produced a number of superior things. Modern science, with its firm grasp of natural reality, is vastly superior to any alternative effort to come to grips with it. The core values of the Enlightenment, with their insistence on the equal value of all human beings, on settling arguments by debate rather than by dogma or force, and on human autonomy and its free deployment in a humane society, are in my view worthy of adoption everywhere. However, value judgements like these are beside the point of the present investigation. In the period 1600–40, no modern science was yet around – it is its bare beginnings that I am seeking to elucidate. Nor was there as yet anything like the Enlightenment, whose very emergence cannot even be conceived of without the prior advent of recognisably modern science. Hence, although the distinctive features and developments of everything that I am about to address carry a label 'made in Europe', none of it should be construed as really meaning 'and therefore worthier than its counterparts in other parts of the pre-modern world'.

In posing here the question of the simultaneous occurrence of those three revolutionary transformations in then current nature-knowledge, I do not *a priori* ascribe the simultaneity to chance. Rather, I set out to investigate which peculiarities in European civilisation may have favoured so remarkable a conjunction. For starters, we take ourselves off to the estate of a Danish castle where on 11 November 1572 at the end of a long workday a small but significant scene was played out.

That evening, Tycho Brahe, returning home to dinner from his uncle's alchemical workshop, noticed a star in the sky where he had never seen a star before. It shone more brightly than Venus and yet this experienced stargazer could not believe his eyes. He first summoned his servants and then some field labourers and all confirmed that it was truly the case – there was now a star where previously there had been none.

What the uneducated found easy to accept without attaching particular importance to it was so far outside the mental framework of academia that initially a scholar might be unable even to see it. The dominance of Aristotle's natural philosophy meant that all change was taken to be confined to the earthly sphere, which includes its atmosphere; beyond the Moon, the universe is perfect and unchanging (consequently, Aristotle regarded comets and meteors as atmospheric phenomena). Novas and supernovas had in fact appeared in the firmament before 1572; Chinese astronomers, obviously ignorant of Aristotle's teachings, carefully recorded their appearance. But if you know beforehand that something is self-evidently impossible, there is a good chance that, when it presents itself nonetheless, you will not even see it.

The fact that now in 1572 a new star was observed in the constellation of Cassiopeia and not just by the youthful Tycho but throughout Europe tells us something about the European situation: there was a new kind of openness. Tycho himself published a pamphlet which concluded, after systematic, sustained and accurate observation, that the phenomenon was indeed taking place outside the Earth's atmosphere. Change was therefore possible in the heavens, too.

That new openness was surely connected with the fact that Aristotle's teachings had lost their monopoly and therefore also their unassailable authority; the old rivals in natural philosophy had returned after a long absence. But there was more going on. Tycho's pamphlet was only one contribution to a tidal wave emerging from Europe's printing presses in which the new star was seen primarily as the harbinger of all manner of disasters. Natural phenomena kept many more people occupied than the few dozen scholars who made them an object of careful study. This widespread interest was mainly due to the 'third' mode of nature-knowledge, that of accurate observation and practical application. This 'third' mode reflected a new openness, an uninhibited curiosity, a readiness to look at things afresh, which had swept over Europe in the wake of the voyages of discovery. Naturally all this openness was only relative; looked at from the vantage point of our modern society in which innovation has become routine, sixteenth-century Europe still looks half-ossified. But that should not be the criterion; the criterion is the totality of civilisations of that time. And, from that viewpoint, Europe distinguished itself by a relatively greater openness and curiosity; a relatively greater dynamism; a relatively greater individualism and outward orientation; a relatively more fervent search for salvation in an active life on earth.

It easily goes wrong when we speak about these things because it is so often done in absolute terms. Take the idea that all Asiatic civilisations were static, inward-looking and without any interest in the outside world. Viewed absolutely, this cannot be maintained. Well before the arrival of the Portuguese, Spaniards, Netherlanders and British there was a thriving inter-regional trade across the Indian Ocean which the Westerners then tried to take over. Indian Buddhism has had a permanent effect on the civilisations of China and Japan. The emperor's Astronomical Bureau in Beijing had a sub-department which was run by Muslims. We cannot therefore speak of wholesale self-sufficiency on the part of the various non-Western civilisations. And yet there is a difference. Europe was less self-contained than any other civilisation; it was restless to a degree not encountered else-where. This was partly due to the relative scarcity of precious metals and luxury items such as silk and spices. Facing a permanent deficit in its international balance of trade, to get what it wanted Europe had to go out and fetch it. The dynamism that this helped to generate found its strongest expression in the voyages of discovery.

But even this kind of activity was not confined to Europe alone. Islamic civilisation also produced some world travellers, who brought back reports about their adventures. Ibn Battuta visited Azerbaijan, East Africa, India and China while our familiar al-Biruni wrote an extensive and to this day still highly readable account of nature-knowledge in India. In the first half of the fifteenth century, the Chinese emperor dispatched fleets to the west under the command of the court eunuch Zheng He. An old dockworker in Mombasa who saw the Portuguese admiral Vasco da Gama's ships at their moorings in the spring of 1498 might have recalled seeing as a little boy Zheng He's much larger junks at anchor in the harbour. The difference is that Zheng He did not continue his voyage and discover Europe whereas Vasco da Gama within a few months discovered India. Another differ-ence is that the imperial court in Beijing soon brought the series of expeditions to an end, whereas Europeans did not cease mapping the world until the last patch of white had been filled in. And finally there was a significant difference in attitude. Chinese sailors tended to display towards the peoples they discovered indifference mixed with benevolent, condescending contempt; their European counterparts a peculiar blend of thirst for profit, bloodlust, an urge to convert and curiosity about the exotic morals and customs that they encountered. One travel report after the other rolled off the European presses. Much of what was observed was naturally coloured by the authors' preconceived ideas and imagination, yet the best of them reflected a genuine interest in the way of life lived by other peoples, which worked its way through society to have a noticeable effect on European attitudes.

That applies particularly to the specifically European mode of nature-knowledge which was directed towards accurate observation. In all kinds of ways it was linked to the voyages of discovery, not just individually as with Admiral João de Castro, but also by way of a collective symbol for everything that this approach to nature stood for. Francis Bacon expressed the connection very clearly:

> Now that in our times the regions of the material globe, namely, the lands, seas and stars, have been opened up and clarified immensely, it would be a disgrace for us if the intellectual globe should remain shut up within the narrow limits of ancient discoveries.

It was the same with the numerous technical inventions which in the Middle Ages helped lighten the burden of field work, crafts practice, warfare and everyday life. Some, notably eyeglasses and the mechanical clock, were homegrown, but many reached Europe from elsewhere. Nevertheless, technical imports such as the halter, the stirrup and the windmill made deeper inroads into European life and were subjected to a more sustained process of renewal than happened elsewhere. The technical enrichment of handicrafts developed its own dynamic which came to be reflected in the 'third' mode of nature-knowledge: nowhere but in Europe did the kind of interface between handicrafts and nature-knowledge develop that we saw beginning in navigation, fortification and linear perspective.

That 'third' form of nature-knowledge had no counterpart in Islamic civilisation. It is true that in the margin of the revival of

'Alexandria' and 'Athens' a few fields of interest emerged that were not Greek in origin but were specific to that civilisation. Whether it was the mathematical problem of orienting prayers towards Mecca or dividing up inheritances, or health care for the community of believers, they were all specifically inspired by the Koran. In Europe there was just one directly comparable instance, the dating of Easter. But in an indirect fashion, the 'third' mode of nature-knowledge does, in its accurate observation and practical orientation, reflect certain specific religious values. This is due to the idiosyncratic direction taken by religious developments in Europe. As the cultural historian and sociologist Max Weber argued more than a century ago, every world religion has an inward-looking aspect and another that is more outward-looking. The ascetic mystic who turns inwards is a fixed feature of every world religion. Unusual about Europe is that in the Middle Ages and the Renaissance the ascetic element is directed increasingly outwards, in the belief that salvation can also and perhaps better be pursued through a frugal life of practical activity. That belief, which is as foreign to the Byzantine variant of Christianity as it is to Islam, was further strengthened in Europe by the peculiar development of the monastic orders and their engagement in enhancing the yield of the land. That relatively strong set of extrovert attitudes also implied taking a certain distance from nature. European Christians do not tend in the first instance to perceive themselves as creatures of nature but, in biblical terms, as stewards of nature with the divinely ordained task to manage it and turn it to their own advantage.

The Reformation reinforced this particular development in monotheism yet further. Not by chance, the great majority of the scholars whose research into nature was mainly empirical and practical were either closely linked to the voyages of discovery or were of some Protestant persuasion (or both of course). In contrast, among 'Alexandrian' or 'Athenian' scholars the proportion between Catholic and Protestant was much the same as across Europe as a whole.

We are now ready to draw two conclusions of special importance for our purposes. The first is that Europe's relatively strong extrovert tendency and dynamism and curiosity find expression in the emergence and eventual revolutionary transformation of the empirical-practical mode of nature-knowledge. The other conclusion is that, in a more general sense, the same peculiarities bring about a climate in which, more than elsewhere, renewal is in the air. It is as if adventurous innovation stands at a certain premium, not in the literal sense of a sum of money, *but in the sense of a value shared civilisation-wide.*

This applies likewise to a final European peculiarity, which is also connected to its extrovert tendencies: an exceptional degree of individualist self-assertion. There are to my knowledge few if any museums in the world with a more varied collection of non-Western art than the Museum of Fine Arts in Boston. You can spend days wandering through the quiet halls and rooms devoted to China, Japan, India, South-East Asia, Islamic civilisation and medieval Europe, before eventually being swept along in the crowds of visitors to the department of Renaissance art. It then strikes you that in those quiet departments, although the styles are of course different, the pattern is nonetheless at bottom the same: sacred art with stereotyped forms that do change over the centuries all the while expressing the same thing, the anonymous artist giving himself over in a self-evident sort of way to a Being or Beings greater than himself. Even where human or animal forms are represented with a certain amount of realism, the sacred character is still unmistakable. In Italy the Renaissance began with a comparable realism, but that soon burst out of its seams, casting the stereotypes aside, and the individual artist reveals himself. Within a few centuries the sacred becomes just one genre among others and is no longer that which gives direction and meaning to whatever is portrayed. Every civilisation encompasses a spectrum that ranges from an awareness of being part of a larger collective whole to a more individual experience of things. In extrovert Europe the individualist end of the spectrum is definitely more prominent than elsewhere.

That individualist self-assertion comes to the fore in the Scientific Revolution, too. Here are three telling passages:

> it being very true that our reputation starts from ourselves, and that he who wants to be esteemed ought to have self-esteem first.
>
> *(Galileo Galilei)*

> if from my youth onward I had been taught all the truths of which I have since sought the proofs ... I would never have acquired this ability and facility I think I have of finding new ones whenever I apply myself to seeking them.
>
> *(René Descartes)*

> Myself, then, I found to be equipped, more than for other things, for the contemplation of truth.
>
> *(Francis Bacon)*

With all their mutual differences, these pioneers of each of the three modes of nature-knowledge which went through revolutionary transformation between 1600 and 1640 were megalomaniacs, utterly convinced of themselves and of their mission. With this trio, the revolution does not remain confined to the already considerable substantive innovations that we have investigated in the present chapter. Their objectives were wider than that; each had a programme to set humanity on the path to a new future. For them, the 're-creation of the world' went well beyond coming up with a new, realist mode of mathematical nature-knowledge, or revealing the mechanisms of particles of matter in motion, or conducting experiments undertaken to lay bare hidden features of nature and to guide thoroughly practical activities like navigation, or curing the sick, in new, unheard-of directions. The thoughts and actions of this particular trio also carried formidable theological, philosophical and political implications.

4   A crisis surmounted

In 1608, Hans Lippershey, a Dutch optician, placed a concave and a convex lens a certain distance apart and enclosed them inside a tube. With the concave lens held up to the eye, distant objects looked larger and nearer, and you could even see things too far away for the naked eye. News of the invention spread quickly and reached Padua in the summer of 1609. The professor of mathematics, Galileo, was one of only two people (Thomas Harriot was the other) to whom it occurred to point the tube with lenses at the skies.

Galileo's idea was far from obvious. We at present take it for granted that the natural world is full of things which are invisible to the naked eye, from the cells in our bodies to the star-studded Milky Way set in infinite space. To observe such things one needs instruments, of a kind that did not then exist. The instruments which did exist were able to support observations and calculations – Tycho Brahe had refined them to the highest degree – but they did nothing but give greater precision to the recorded properties of objects already known. Nobody could have guessed that the Milky Way, that misty veil lying across the night sky, would on closer inspection dissolve into millions of stars. Nobody could have suspected that Jupiter is orbited by moons, that there are strange appendages on both sides of Saturn or that the surface of the Moon is studded with craters and valleys. All these facts were discovered by Galileo, and their repercussions would extend much further than the immediate sensation they caused throughout Europe when he published them in 1610 in a concise, matter-of-fact and spectacularly illustrated treatise entitled *Sidereus nuncius* ('The Starry Messenger'). They were also to have a number of important repercussions for Galileo himself.

In the first place, they gave him the opportunity to get away from Padua. The previous eighteen years of experimenting, reasoning and checking had laid the foundations for a radically new mode of realist-mathematical nature-knowledge. He was entirely convinced of its superiority over any current philosophy of nature. He now saw himself as a 'mathematical philosopher'. Not that this was a recognised social role: there were mathematicians and there were philosophers, but in between yawned a wide gulf. Galileo's philosophical colleagues did not know what to make of it – after all, mathematics and the real world had from the early Alexandrians onwards been wide apart. It was precisely because philosophers dealt with the real world that they were paid so much more than mathematicians with their causally empty calculations on fictional models. Galileo was always keen to prove that he himself knew best, but his colleagues shared so little common ground with this 'mathematical philosophy' of his own making that they would not even give him a chance to prove its superiority in open debate. He also took exception to his far lower salary. And so this born and bred Florentine seized on his telescopic observations and informed the Grand Duke of Tuscany that in his forthcoming publication he intended to name Jupiter's moons after his family. After extensive negotiation, Cosimo II de Medici accepted the offer, with the result that *Sidereus nuncius* not only contains a dedication to the Grand Duke, full of baroque sycophancy, but also regular references to Jupiter's moons as the 'Medicean stars'.

In exchange, Galileo returned to his home town as court mathematician in the service of the Grand Duke. And not merely as court mathematician, for he made a special point of having his function officially described as 'philosopher and mathematician'. He was now on equal footing with the philosophers who from now on would have to engage him in debate. He too was now a philosopher, even if of a type as yet unknown, who now enjoyed the support of the court and also a much higher income.

NATURE-KNOWLEDGE AND RELIGIOUS
CONCEPTIONS OF THE WORLD

When, in 1611, Galileo visited Rome, he was certainly given a hearing; in fact, the visit turned into a triumphal procession. He was received by the Jesuit fathers of the Collegio Romano, skilful astronomers who, led by the aged Clavius, repeated and confirmed his telescopic observations. This was important for Galileo, who himself was not entirely clear how the instrument actually worked – not unreasonably, some even refused to look through an instrument which apparently operated by magic and illusion. But the public support he received from Clavius and also from Kepler, the Imperial Mathematician – that counted.

Meanwhile, exchanges with the philosophers at his old university in Pisa were not going well. Characteristically, they involved a problem on which Archimedes and Aristotle had expressed widely differing opinions without the discrepancy ever giving rise to debate. Archimedes had used abstract mathematics to derive the conditions under which an object placed in a liquid floats or sinks. These were incompatible with Aristotle's views on heaviness and lightness, and on floating in connection with the shape of an object. But now, in 1611, on the occasion of one innocent question, the two accounts collided. At a rather smart luncheon party, some nobleman wanted to know how it is that ice floats on water. To this the mathematical court philosopher gave a very different answer from that of the university philosophers. Galileo applied Archimedes' propositions to reality and attributed the phenomenon to a difference in specific gravity — freezing makes water less dense and therefore lighter. His Aristotelian opponent attributed the phenomenon to the shape of the ice. Moreover, he thought that he could hoist Galileo by his own petard. Was Galileo not in favour of experimental testing? Well, ebony is clearly heavier than water so, if Galileo was right, a slice of ebony should sink. How would the test turn out?

For both sides, there was a great deal at stake.

If Galileo was right, and the slice of ebony sank resolutely to the bottom, Aristotle's entire edifice would in effect come crashing down. After all, if specific gravity is the decisive factor, then heaviness and lightness are no longer absolute opposites as in Aristotle, but merely a question of more or less heavy. The whole idea of movement as oriented to an object's natural place would collapse, and with it Aristotle's entire account of motion as such, and with it his conception of change as the realisation of an immanent end, and with it the very core of his natural philosophy. Precisely that which gave Aristotle's philosophy its extraordinary power, to wit, its tight inner coherence and the way it seemed to reflect everyday experience, now threatened to turn on itself – pull one beam out of the tightly constructed building and it collapses in ruins. Aristotelians were well schooled in logic and they saw the danger all too clearly. The pent-up anger of their polemics against Galileo can only partly be explained by what they found so bizarre in his manner of connecting mathematics to reality – in fact, everything that still makes the knowledge structure of mathematical science with its three levels of reality so hard to grasp. The professors further realised all too well that they had hung not only their intellectual souls but also their source of income on the peg of a collection of doctrines whose validity now seemed to depend on the behaviour of one slice of ebony in water. What a relief when it stayed afloat!

Its floating was bad luck for Galileo. The fact that a slice of ebony, despite its higher specific gravity, does not sink is due to something that he was not and could not have been aware of, namely the surface tension of water. And so he had to put a great deal of effort into rescuing his position with somewhat tortuous reasoning. More to the point, how could he now put across his fundamental belief in mathematics as the key to explaining nature when at the first attempt he had been forced on to the defensive? In all the hullaballoo he was fortunate that his new status at court allowed him from a certain point onwards to leave the professorial gentlemen to stew in their own juices. Thanks to Jupiter's moons the tables had been well and truly turned.

The somewhat comic mock battle of 1611/12 is what must have persuaded Galileo from then on to present his realist-mathematical mode of nature-knowledge differently. For that he reached back to Copernicus. It is not known precisely when he became an adherent of Copernicus and what persuaded him that the Earth turns on its axis and revolves around the Sun. In 1597, in a reply to Kepler's first letter to him, he wrote that he too was a Copernican but that he wanted to keep it to himself. He continued to do so until 1612. At least three reasons eventually led him to step forward as the great defender of Copernicus, and to launch a campaign to persuade as many people as possible, and particularly the Catholic Church, to accept the dual motion of the Earth.

One reason was his telescopic observations. Before them there had been no empirical evidence for the Earth's motion whatsoever. Supporters of the idea might point out the greater coherence and simplicity that it lent to the planetary system when compared with Ptolemy's model of the universe. They might come up with arguments from natural philosophy or try to undermine the obvious common-sense objections. But that was all reasoning. There was just no factual evidence. Galileo's telescopic observations put an end to that state of affairs for good and, as well as earning him his court appointment in Florence, it is here that we encounter the second major consequence of his observations. The Moon reveals craters and valleys, showing it to be more like Earth than a perfect crystal sphere where nothing ever changes. Jupiter has moons, just like the Earth, so the possession of satellites does not necessarily indicate a non-planetary status. The Sun, Galileo soon discovered, has spots, which means that it too is not perfect. And, to cap it all, at the end of 1610 he observed that the planet Venus, just like the Moon, shows phases, which is squarely ruled out in Ptolemy's system. In short, Galileo could now boast some additional, powerful pieces of evidence. He even thought that they were sufficient to prove Copernicus completely right, though in this he was mistaken – with catastrophic consequences.

A second reason for starting a campaign in defence of Copernicus was the affair of the ice and the slice of ebony. Once the dust had settled, Galileo decided that the Earth's motion was a much more appropriate vehicle for the realist-mathematical mode of nature-knowledge which, all considered, was his primary concern. Due to its complex knowledge structure, it seemed fated to run into obstacles which only a fellow expert could overcome. But Copernicus' planetary system, certainly in the simplified form of Book I, and enriched with Galileo's new conception of motion, embodied his new idea of mathematics as the key to the natural world in a way that any layman with an academic education should be able to follow.

And so for the first time in the seventy years since Copernicus had died and his book had been published, those philosophical and religious issues rose to prominence which the author himself had so carefully circumvented. Copernicus had managed to ignore that the dual motion of the Earth (on its axis and around the Sun) represented a radical break with Aristotelian and indeed every other form of natural philosophy. When the Earth is no longer the immovable centre of the universe but is just one of six planets orbiting the Sun, what remains of the dichotomy between things in unceasing flux on Earth, in the sea and in the atmosphere, and the perfect, unchanging world beyond the Moon? Even worse, the entire universe was in danger of coming apart at the seams. Copernicus had already noted that, if indeed the Earth goes round the Sun in a year, a person observing a star with an angle gauge ought to get slightly different readings in the spring and autumn, the Earth then being on the other side of the Sun. But no such angular difference manifested itself. Copernicus explained its absence away on grounds which, however valid in retrospect, could not but look quite weak at the time – the distance from the Earth to a star is just too great to make the difference visible. That is, the space between the sphere of Saturn and the fixed stars is so vast that viewed from a star the whole orbit of the Earth is no larger than a dot. This space served no other purpose than to dig Copernicus out of a hole. For this and other reasons, Tycho had rejected the idea of the Earth's motion. But it was also possible to argue the other way. If the Earth is actually a planet, it opens up the prospect of a universe that is not merely enormously enlarged but, who knows, perhaps even infinite. At the end of the sixteenth century, Giordano Bruno, a monk whose world-view had strongly magical overtones, coolly drew that conclusion. He even announced that that infinite universe is filled with solar systems just like our own. Not that Galileo shared those views; he never ceased to regard the universe as finite. But the important point is that anyone who argued for the actual, not just fictional, motion of the Earth, be it on reasoned grounds or from observation, was opening a veritable Pandora's box.

By 1613, the Grand Duke's family had begun to suspect as much. After all, the family's name was attached to the moons of Jupiter which were now being used to question whether the Earth really is the immovable centre of the universe. How did it all fit in with the Bible story of Joshua, the Medici asked Galileo. Joshua, Moses' successor, was busily relieving a besieged city when the Sun was about to set. God performed a miracle and made the Sun stand still, thus giving Joshua welcome time to complete putting the enemy to the sword. Of course, such a miracle is possible only if the Sun ordinarily revolves around the Earth and not the other way round. There are other passages in the Bible which either state or imply that the Earth stands still. Could Copernicus' supposition be reconciled with Holy Scripture?

A similar question had arisen in the fourth century in connection with the spherical shape of the Earth. The most authoritative of the Church Fathers, Augustine, had responded that Holy Scripture has not been set up as an astronomy textbook. Those who wrote down the books of the Bible had adapted their language to current conceptions, which was not really a problem since our salvation does not depend on the shape of the Earth. Passages implying a flat Earth need not therefore be taken literally.

This was the kind of interpretation that Galileo appealed to when he replied to the Grand Ducal family a few days later in a semi-public letter. In it he acknowledged that only the Church had the authority to decide whether certain passages in Scripture should be interpreted literally or not. He knew very well that it was not a straightforward matter: you cannot tinker with the Word of God as if it were a trifling business. Nevertheless, he argued, the guardians of Church doctrine would be well advised to consult the experts, the only people able to decide what the position and the physical qualities of Earth and the heavens really and truly are like. And who were these experts? They were the mathematical philosophers or, in modern parlance, the natural scientists. Once they had demonstrated that the Earth revolves around the Sun, the Church should not keep identifying itself with an outdated standpoint. If it did so, it would be putting at risk the whole doctrine of salvation and its source in the revealed Word of God.

Galileo was a faithful son of the Catholic Church and there is not the slightest reason to doubt the sincerity of his beliefs. Furthermore, the four centuries that have followed these events show that he indeed perceived this lethal danger to his mother Church with utmost clarity. Even so, his argument unmistakably implies a claim to the power of final decision. The theologian is being demoted to carrying out what only the 'mathematical philosopher' can pronounce final judgement on, the question whether in a given case God's Word should be interpreted literally or not. And the profession of mathematical philosophers counted as yet no more than just a single representative.

It is not strange that when the Florentine clergy got wind of Galileo's 'Letter to the Grand Duchess' they immediately lodged a protest in Rome. It is also not surprising that Cardinal Bellarmine, to whom the pope had entrusted the guardianship of Church doctrine, challenged Galileo to come forward with proof. For astronomers to treat the Earth's motion in the customary way as a useful fictional model was unproblematic. But to start reinterpreting the Bible without clinching proof that the Earth truly orbits the Sun was quite out of the question.

We have meanwhile reached the year of our Lord 1615. At this point, Galileo might still have stepped back. Roberto Bellarmino SJ was one of the most senior members of the Inquisition. As such, he had had a hand in condemning Giordano Bruno to be burnt at the stake (although this had nothing to do with Bruno's ideas about an infinity of solar systems). Via the Jesuits of the Collegio Romano, Bellarmine was also well aware of both the merits of, and the objections against, the idea of a moving Earth. Why couldn't Galileo be sensible and settle for the unproblematic treatment of the Earth's motion as a fictional model? But Galileo's whole campaign was based on its being real, not fictional, for only thus could he spread his core conviction of the mathematical nature of reality. To stop that campaign now was not an option. Meanwhile both men were in fundamental agreement on the need for caution in interpreting the Bible. What remained was the question of 'proof' for the dual motion of the Earth. At this point Galileo might have been well advised to take an intermediate step and find out what Bellarmine actually meant by 'proof'. The arguments for the Earth's rotation were not necessarily weaker than those earlier ones for a spherical Earth, so why not invite the cardinal politely to treat them in the same way? Galileo knew perfectly well that opinions in the Church about proper Bible exegesis diverged to some extent. By no means everyone at the top of the hierarchy shared the lower clergy's penchant for taking every word in the Bible literally. In short, the situation was precarious and the nearer that Galileo could come to the precedent of the spherical earth the better it would be.

The tragedy of what happened in 1615/16 is that neither side behaved with the caution that the situation required. Galileo was convinced that he had irrefutable proof. He thought – wrongly – that the regularity of the tides is due to nothing but the daily and annual motions of the Earth combined. Against the advice of the Tuscan ambassador in Rome and Cardinal Bellarmine, who knew their pope, he set off for Rome to persuade the Church that he was right. For months on end, he went from one cardinal's court to the next.

He argued, debated, and with his verbal brilliance and arrogant display of superiority made admirers and bitter enemies alike. He even succeeded, without consulting Bellarmine, in having an intermediary bring the matter up with the pope himself who, unfortunately, was a career diplomat deeply mistrustful of new-fangled scholarship. Shocked by Galileo's demand that the Church should adopt the proposition of a rotating Earth, he set up a commission of enquiry. And within a week, as in a classical tragedy, Galileo had brought about the very opposite of what he had set out to achieve. It was Galileo's campaign that finally moved the Church to take up a position on Copernicus after seventy years, and that position went directly against what Galileo had so insistently advocated. For the first time, there was now a decree, although in its undue haste very clumsily formulated, in which the dual motion of the Earth was strongly condemned as nonsensical and in conflict with Holy Scripture.

As yet the effects of the decree were rather limited. Galileo had to promise not to spread Copernican doctrine, but this was done privately and the public remained unaware of it. Also Copernicus' book appeared on the Index of Prohibited Books. But that was about it. Perhaps the greatest setback was that the Jesuits were now bound, too. Phases of Venus or no phases of Venus, they were trapped by an immovable Earth. As far afield as China, their work would suffer from this.

Seven years later a friend and admirer of Galileo's was elected to the papacy. Six audiences with the new pope, Urban VIII, sufficed to persuade Galileo that he was now free to reopen his campaign provided he gave the impression that he did not put forward the motion of the Earth as something absolutely certain but merely as a useful supposition. After all, so the new pope insisted, God in his omnipotence might have arranged things in wholly different ways quite unfathomable to human understanding.

And so Galileo resumed his campaign. In his *Dialogo* he personified himself as the defender of a simplified version of Copernicus' system, and the Aristotelian world-view was defended by a caricature simpleton, while the third participant played the part of the intelligent layman who asks the shrewd questions. It is both a scientific and a literary masterpiece; its depth is clad in wit, and prior familiarity with the subject matter is hardly necessary. On the first of four 'Days' Galileo attacks the Aristotelian conception of the world with the help chiefly of his telescopic observations. On the second Day he refutes objections to the Earth's daily rotation, and that is where he develops his new conception of motion and how it is retained. On the third Day he tackles objections to the annual revolution around the Sun. On the fourth Day he makes his treatise on the tides decide the matter. Just before the end, he manages to put the pope's favourite argument into the mouth of none other than the Aristotelian simpleton who then persuades his partners in debate, struck dumb by this 'wonderful and truly angelic doctrine', seemingly to adopt it, as it were.

The pope was furious. He set up a commission of enquiry which faithfully reported back that Galileo had only pretended to present even-handedly the arguments for and against a moving Earth. In reality, the *Dialogo* was no more than a persistent defence of Copernicus' idea as realistic and true, not as fictional. Galileo had undeniably broken the promise he had been required to make in 1616. And so in 1633 it came to the scandal of the trial. Placed under psychological pressure by the Inquisition (he was never tortured), the 69-year-old Galileo renounced his Copernican conviction. The decree of 1616 now became public, but the spectacular nature of the trial itself, and the wide distribution of its official report, amply sufficed to make it unavoidably binding on everyone who fell under the authority of the Church.

The consequences were immense. True, for Galileo personally it could have been worse. His punishment went no further than house arrest in a village outside Florence and daily psalm recitations (soon taken over by his daughter, a nun). But publishing anything in Italy was now out of the question. He managed to have the *Discorsi* smuggled out of the country and published in the Protestant Northern Netherlands with its much greater freedom of the press. And even the offensive *Dialogo* was published by a Protestant publisher within two years of the trial in a Latin translation. Meanwhile, in Italy research dried up to a large extent; only investigations devoid of philosophical-theological implications had a chance of being permitted. Outside Italy, too, the trial proved consequential. In 1633 Descartes was well on his way to writing down his philosophy of moving particles in a manuscript with the modest title *Le Monde* ('The World'), when he heard of Galileo's trial and the verdict. He promptly hid the uncompleted manuscript away in a drawer which it did not leave (apart from access given to the occasional disciple) until after his death. Descartes, like Galileo, was Catholic but lived and worked in the Dutch Republic where at least in a direct sense he had nothing to fear from the Inquisition. So why the self-censorship?

Descartes was all too well aware that the philosophical-religious implications of his own work, too, were considerable. The world that he sketched in the book of that name was infinite and full of solar systems like ours. We are already familiar with his starting point: God created the world in the form of particles of matter, which while unceasingly forming whirlpools move according to fixed laws. In *Le Monde*, he showed how this applied universe-wide, smoothly incorporating the telescopic observations of Galileo and other more recent observers. Underlying all this was his idea that there are two substances in the world, 'extended substance' (*res extensa*) and 'thinking substance' (*res cogitans*). The latter substance is reserved exclusively for human beings. Aristotle and other natural philosophers had distinguished between three 'souls'. Plants have a vegetable soul, animals have in addition an animal soul and humans a rational soul as well. Descartes did away with the first two; plants and animals now came under the category of 'extended substance'. And even humans fell to a large extent under it. Not only are plants and animals compounds of particles, but the physical functions of the human body run by the same machinery. In *Le Monde* Descartes devoted a lengthy chapter to a detailed description of 'Man' as a machine to which God has attached a portion of thinking substance, that is, our immortal soul.

After hiding away his uncompleted debut, Descartes became engrossed by the problem of how to publicise his conception of the world after all. He attached great importance to it because his deepest ambition was to be recognised as the new Aristotle. He was convinced that his natural philosophy, in contrast to all others, was the only true one because he was the only one to have built it up from completely certain, truly and absolutely indisputable foundations. Of course, the competitors of old had all claimed the same, but even back then the Sceptics had shown that to be impossible. Powerless to counter the Sceptics' assault on the foundations themselves, the Platonists, Aristotelians, Stoics and atomists had taken the easy way out, acting as if it had never happened. So the Sceptical dragon still had to be slain. To do this our St George girded himself with a metaphysical treatise, the *Meditations*, which after the setback of 'The World' he saw as an indispensable intermediate step towards piloting his central message safely into harbour.

The Sceptics' critique, we recall, amounted to an extensive argument that our intellect and our senses are in the habit of deceiving us in countless ways. We can never be certain of the conclusions that our intellect makes us draw or of the information with which our senses provide us. The only thing we can do is to suspend judgement. Descartes' master-stroke consisted in taking this critique completely on board and then going one step further than any Sceptic had ever done. He conceived of the famous *'malin génie'*, an evil demon who is deliberately out to mislead us. Under this blistering attack of sceptical doubt, everything goes up in smoke; the whole world and everything in it has been thought away and nothing is left. Nothing at all? Yes, there is still something. Even while I doubt everything, I cannot avoid being aware that those doubts are being thought by me. *Cogito ergo sum*: 'I think therefore I am.' Even the most dyed-in-the-wool sceptic cannot wriggle out of that. I cannot simply think my doubting self out of existence. And so after all a bedrock of incontrovertible certainty has been found. The next task then is to build bridges from this unshakeable rock back to the world. However, it has now become a different world, no longer the world of everyday observation but the world of extended and thinking substance. For Descartes, the route back to that world runs via mathematics.

Mathematics was of special significance for Descartes in three respects. He himself was a mathematical genius, one of the great innovators. His contribution was an enormous step forward toward the identification of geometry and algebra. He took that step in a treatise entitled *Géométrie*, which was published in 1637 in one volume with studies of refraction and of atmospheric phenomena. The introduction to these three 'essays' was a brief announcement of his natural philosophy under the title *Discours de la méthode*. This collection formed Descartes' debut publication, meant to test the waters in the tricky circumstances that Galileo's condemnation had created. Two of the treatises showed what Descartes was capable of in the mathematical mode of nature-knowledge. He did not cultivate it in the revolutionary manner of Galileo, but in classical Alexandrian style. The law of refraction which Ptolemy had been unable to find but Ibn Sahl had discovered, as too had Harriot and Snel, was now first published by Descartes. But like Ibn Sina and others before him, Descartes' philosophical and mathematical strands of nature-knowledge were kept separate and had hardly any impact on each other. Indeed, they actually contradicted each other on some points which Descartes did not even bother to resolve. Meanwhile, mathematics fulfilled a third role in his Athens-plus natural philosophy: it was the prime example of incontrovertible, certain knowledge.

When he was first taught mathematics in a Jesuit college, he was already impressed by the complete certainty with which you proceed from one theorem to another. But how is it possible? What makes mathematics such an infallible vehicle of certainty? Years later, Descartes concluded that it is the 'clear and distinct' line of reasoning that characterises mathematics. If he therefore were to reason clearly and distinctly from the bedrock of certainty contained in his 'cogito', he could with incontrovertible certainty deduce the entire world from his intellect. Only when, in this top-down process, he reached the lowest level of individual natural phenomena would any room be left for uncertainty, as God might have chosen to arrange for some detail in this way or in that, so that only empirical observation could show conclusively how He had arranged it in fact. For the rest, the world lay firmly and incontrovertibly fixed. It was a remarkable conclusion to an intellectual exercise that, in the opening passage of the *Meditationes*, had started with a rousing call to every reader to ask himself at least once in his lifetime whether everything that he has always assumed to be true actually is true.

Having laid these foundations, Descartes felt that only one more thing was now needed before he could go public with his complete natural philosophy. The dual motion of the Earth had been condemned by his mother Church and he had no intention whatever of crossing her in any way, but now he reckoned that it was not really necessary. He could circumvent the ban by proposing that it is not the Earth itself that moves but only the vortex which drags it through the solar system. And Descartes would not have been Descartes if he had not been able to think up an appropriate piece of reasoning to accompany it. And so the path was now clear for the publication in 1644 of his chief treatise, which he proudly entitled *Principia philosophiae*.

All the same, he had meanwhile been caught up in a conflict of his own with an ecclesiastical organisation. The conflict was not with his own Church but with Gijsbert Voet (pronounced 'Voot'), an extreme Calvinist minister. Voet was the leader of the 'Extended Reformation', a movement that aimed to turn the Dutch Republic into a theocracy, where all other religious denominations would be banned and the political elite, the 'regents', would be downgraded merely to carrying out such policies as the church ministers had agreed on in a kind of seventeenth-century ayatollah regime. Furthermore, Voet was the vice-chancellor of the newly founded university of Utrecht, the city in which Descartes was then living. It was rather a sham fight because a clever disciple of Descartes, Professor Henrick de Roy ('Regius'), went into battle on his behalf and took most of the blows. For Descartes the episode remained confined to a battle of words, fought out in fishwives' style by both sides, which fizzled out in the end. Voet managed to get the Utrecht city council behind him, yet its authority did not stretch far enough to do Descartes serious harm. To be on the safe side, the French ambassador put in a good word with the *stadhouder*, Prince Frederik Hendrik. The significance of the episode lies elsewhere. It was both a part and a symptom of a crisis of legitimacy which overtook both Alexandria-plus and Athens-plus on the European continent in the 1640s.

A CRISIS OF LEGITIMACY

It goes almost without saying that the innovations that, notably, Galileo and Descartes proposed and backed up with at times fiery and often powerful arguments attracted enthusiastic support as well as arousing determined opposition. In the end the support would win the day, and indeed overwhelmingly. Modern science, for which they provided numerous initial impulses, has survived and has even managed to multiply and accelerate to this day. But not during the period immediately following the pioneers. Between about 1645 and 1660, the fate of these new ideas hung by a thread, and we must now investigate how and why this happened.

Let us first look at the fears which the innovators, Descartes in particular but also Galileo and even Kepler, aroused in Professor Gisbertus Voetius:

> The essence and existence of [substantial] forms are being denied ...
> Once this dangerous axiom is granted, the vanity, scepticism and
> excessive licence of the human mind shall lead it down the slope of
> adversely arguing that there is no rational soul, no generation and
> conception of man in the mother's womb, no wind, no light, no
> Trinity, no Incarnation, no original sin, no miracles, no prophecies,
> no awakening of a sense of God in the human mind and will, no
> regeneration of man through God's grace, no demonic action inside
> the body of man or around his mind.

In short, if you take the heart out of Aristotle's teachings you condemn the world to incomprehensibility and rob Christianity of its principal doctrines.

How is that possible? How could an admittedly hair-splitting yet highly intelligent, erudite and theologically expert scholar like Voet assert without blushing that not only the intelligibility of the natural world depends on Aristotle (not a strange view at the time) but also the very foundations of Christian faith?

The explanation lies in the alliance between Aristotle and Jesus Christ which Thomas Aquinas had hammered out three centuries earlier. That alliance was not confined to the abstract level of God's omnipotence which Thomas had so cleverly reconciled with Aristotle's aptness to regard a phenomenon as insufficiently explained until it is clear why it cannot possibly be other than what it is. Particularly in Roman Catholic doctrine there is a very close connection in the central part of the mass where bread and wine are actually transformed into Christ's body and blood. It is not by chance that this is called 'trans-substantiation', a term full of Aristotelian overtones. It is the substance itself which changes, not its attributes. The obvious question was how that could be fitted into Descartes' natural philosophy. It might of course be argued that it is irrelevant because revealed truths are not in need of proof. But throughout the history of religion, and particularly the monotheistic religions, things have not worked out that way. Without evidence acceptable to the intellect it was hard indeed to win over more than just a few among the educated.

In consequence, the connection in Christian Europe between natural philosophy and faith was very close indeed and an attack on the former could easily be regarded as a threat to the latter. And now the attack came from two sides. On the one hand, there was the strange 'mathematical philosophy' which Galileo thought should replace current natural philosophy under the absurd banner of the dual motion of the Earth. On the other hand, there was this natural philosophy which Descartes was presenting as the wholly new, incontestably certain alternative for all the old ones.

And it went much further than merely undermining the support that Christian beliefs received from current natural philosophy.

There was the exegetical question of how to read Holy Scripture. Did it at all times and everywhere have to be understood literally or not? We have already explored this question in connection with Galileo's conflict with the Catholic Church.

There was also the question of our immortal souls, which was closely linked to the essential difference between man and beast. Conservatives are sometimes more aware than the innovators themselves where their innovations may lead, whether intentionally or not. Descartes never understood why Voet and his ilk should doubt his piety. Had he not in his *Meditationes* even produced a new, incontrovertible proof of God's existence? And had he not with his dichotomy of extended and thinking substance defended our immortal souls by separating them off as precisely as possible from the irrational animals? Voet saw much more clearly than Descartes himself that very little effort was now required to take that thinking substance, so loosely and obscurely attached to our bodies, and cut it out of the universe with one final stroke. Unintentionally and probably without quite realising it, Descartes had come close to a purely materialist conception of the world in which there would no longer be any place left for an immortal soul or for God's sovereign action. Descartes was not Spinoza, but in a manner of speaking Voet saw him coming – inevitably, at some point someone would draw the pantheist or atheist implications from Descartes' doctrine which he himself seemed to have overlooked.

Something similar was the case with Descartes' process of systematic doubt. He himself felt that he had left the Sceptics no conceivable way out. But again no more than a small step was needed to put even his *cogito* through the sceptical mill in its turn. Thus Descartes appeared to herald the resurrection of a reinforced and therefore even more dangerous scepticism. And what about his recommendation to doubt, at least once in your life, everything you have ever taken to be true? Where might that lead to? That all of us begin to think for ourselves? To be sure, Descartes himself had quickly done an about-turn and resolved any radical doubt in the new certainty of his natural philosophy. But for others he had opened the way to denying any form of authority, and in particular intellectual authority. What, for God's sake, was the world coming to?

Come to that, what kind of world was it that Descartes had sketched in his natural philosophy? The cosmos had taken on the guise of a clock, progressing mechanically with pitiless regularity and subject to uncompromising natural laws. What was left of us humans, those minuscule pieces of thinking substance in an infinite universe, chock-full of randomly swirling particles of matter?

It is true that Voet's fight with Descartes did not directly achieve much. Descartes as an individual remained unharmed, and publication of his books was not obstructed. Even for Galileo, as we saw, the consequences had been fairly limited; he lived out the last nine years of his long life under house arrest but his work remained accessible to the public, even if hardly in his native country.

Let us take a closer look at the course that these two conflicts took, and in particular at their 'end games'. Then it turns out that these battles were not fought out over the heated controversial philosophical and theological issues that we have been examining, but that they were poured into a judicial mould. The whole trial against Galileo turned on the question whether in the *Dialogo* he had broken the injunction of sixteen years earlier not to defend in public the dual motion of the Earth. And in Utrecht the tone was soon set by trivial issues such as the true authorship of a certain pamphlet. Furthermore, secular authorities had an important say in the outcome of both affairs. The Inquisition was of course an ecclesiastical institution, but the pope was also the ruler of the city-state of Rome and as such a colleague of Galileo's patron the Grand Duke of Tuscany; this imposed certain limits on what he could do with the duke's client. And in the Dutch Republic not only the Utrecht councillors but also the regents elsewhere were, to Descartes' disappointment, more concerned with maintaining peace and quiet than with stamping his opponent underfoot. In 1657 his re-interment in France (he died in Sweden in 1650) again gave rise to a similar conflict, which dragged on for decades. In the end, his supporters even got into trouble with the king and with the archbishop of Paris, and again we see the same pattern: the case shifted from the philosophical and theological issues which lay at the heart of the dispute to juridical hair-splitting. But here, too, those giving offence faced nothing more serious, if things came to the worst, than being dismissed from a professorship. There was no question of imprisonment, let alone execution. And again the books containing the offending passages continued to be published, sometimes with a few symbolic corrections, if not in France then in the United Provinces.

What was actually going on here? Why was this kind of smoke screen being raised time and again?

In the first place, all this reflected the many divisions that ran across Europe at the time. The subcontinent was divided up into numerous sovereign states, governed by princes who were not at the same time religious leaders. They had come to accept a certain division of power with the ecclesiastical authorities. In a universal empire united under one single, secular *and* spiritual sovereign answerable to no one but himself, a Galileo or a Descartes would in all likelihood have been given short shrift to the point of being summarily executed. But in Europe the pioneers survived their offensive opinions which in any case continued to appear in book form. Is it then not vastly exaggerating to speak solemnly of a 'crisis of legitimacy'? Can it really be maintained that in such circumstances the revolutionary new modes of nature-knowledge were in danger of losing their anchoring in society? Events of a quite different nature had brought about a crisis of legitimacy in Islamic civilisation around 1050. Driven by a chain of invasions, Islamic civilisation had turned in on itself, leaving for centuries hardly any room for the further pursuit of nature-knowledge. In Europe, did the offence that the new ideas gave to many have comparable consequences?

Indeed such an outcome was quite near the horizon. The conflicts sketched above were really far from being storms in teacups; their consequences extended much further than the relative impunity of the main characters and their publications might suggest. It is not for nothing that the years between about 1645 and 1660 were the least productive period of innovative nature-knowledge during the seventeenth century taken as a whole. Only in retrospect do those years look like a brief lull. Viewed at the time, it looks as if momentum slackens and revolutionary innovation comes to a creaking halt. But was it really due to rapidly declining legitimacy, that is, to a widely shared sense that the views promulgated by Galileo, Descartes and other innovators were so bizarre and so sacrilegious that no decent society could tolerate them?

The question can be answered with the help of three key concepts: censorship, self-censorship and the threat of war of all against all.

We have just noted some of the positive effects of the many divisions in Europe. Both within and between European countries there were so many different forms and gradations of power-sharing and conflicting authorities that the new, offensive ideas could not simply be beheaded in a single blow. However, just at this time these divisions of power and mutual hostility almost amounted to Europe-wide civil war. To be more precise, ever since the Reformation in the early sixteenth century, Europe had been plagued by wars of religion. Closely linked to the Habsburg attempt to impose a continent-wide hegemony, things began to get completely out of hand with the Thirty Years War. That conflict, which divided and laid waste to Germany between 1618 and 1648, had drawn in all the other great powers, each of which faced domestic revolts of their own that, in England's case, even amounted to a genuine civil war. In the 1640s there was a real threat of a war of all against all and of Europe sinking into chaos and anarchy.

For the two centuries since the Greek legacy had been revived in the middle of the fifteenth century, the centres of pursuit of nature-knowledge had been in Italy, Austria and southern Germany, France,

England, the Netherlands and to a more limited extent the Iberian peninsula. We shall now trace how nature-knowledge was affected in each of these centres.

About Austria and southern Germany we can be brief. The Thirty Years War knocked out the entire territory. It would be many years after the war had ended before there was once again any cultural activity worth mentioning.

In Italy, after the trial of Galileo the Inquisition reacted sharply against anything that might cause offence. The decree of 1616 had already removed from the Jesuits any room for the relatively uninhibited pursuit of their studies. Galileo had barely managed with artful juggling to get his fiery, witty defence of the Earth's dual motion published, with the fatal result of the trial in 1633. From then on, the only way left to pursue Galileo's revolutionary innovations any further was through either highly abstract mathematical investigations or wholly practice-oriented ones. As soon as any ideological consequences could be detected, intervention by the authorities was guaranteed. Naturally there were always bypasses and roundabouts, but that way you lose so easily the unfettered openness of mind which true innovation cannot do without.

All of this not only excluded Italy as a further breeding ground of innovation but also all the other regions where the Inquisition was in a position to rein in people's thinking and its outcomes. It applied to Spain and Portugal and also to the Spanish Netherlands. One of the great pioneers, van Helmont, experienced it personally. He was imprisoned on several occasions, and the lengthy work in which he recorded his radical renovation of Paracelsus' doctrines was only published four years after his death, and not in his native country.

In contrast, in England the publications of the pioneers Bacon, Gilbert and Harvey had given no offence. But although the civil war that had led to Charles I's execution had also led to the collapse of censorship, the accompanying troubles did not allow much opportunity to pursue and expand their pioneering work in a consistently constructive vein.

That left only the Northern Netherlands and France as possible flag-bearers of innovation. And this is precisely what gave the two conflicts surrounding Descartes and his posthumous partisans their historical significance. In what way did these conflicts affect chances of further innovation?

In the Dutch Republic the situation was not encouraging in this regard. Everywhere else there were princely courts that could seek to provide patronage and often did, but in the Republic there was no such court. In the cities, intellectual interest in nature-knowledge was pretty well limited to activities of direct use for trade and navigation. In the universities, Aristotelianism prevailed. Descartes had fastened his hopes on these universities; part of the motivation for his settling in the Republic was the hope that here he might start his career as the new Aristotle. The conflict with Voet nipped this hope in the bud. In Utrecht, except for a brief interlude, Descartes' doctrines did not stand a chance until well into the next century. In Leiden, after numerous quarrels, a compromise solution was found in the form of a cautious mixture of Aristotelian and Cartesian doctrines. The end result was carefully stripped of its ideological spines and embellished with the title of 'new-old philosophy'. This type of half-baked solution was a long way from promoting any ongoing revolutionary renewal.

So that seems to leave France as the sole hope and refuge for genuine innovation.

In the period which we are discussing there was little indication of this. It is true that the Inquisition had much less of a say there than in Italy and the Spanish territories, but it could certainly not be ignored. It did make a difference that in the course of the conflict over Descartes' intellectual legacy, a French Jesuit succeeded in having a wide selection of Descartes' works placed on the Index of Prohibited Books. Incidentally, it is revealing that on that list which for the most part was circumscribed only loosely, we encounter the two pamphlets that Descartes had written against Voet. So the man who aims to establish a Calvinist theocracy in the Dutch Republic ends up being protected by Rome against the attacks of a fellow Catholic! It is a detail that indicates just how much Voet's concerns about Descartes' ideas were shared by a far wider circle than just his own supporters in just his own country.

The continuing war of words over Descartes' inheritance left other visible traces behind. As well as Beeckman and Descartes, there was a third scholar who had made an effort to construct a world of particles in incessant motion. That was Pierre Gassendi, a French priest whose piety was often questioned. He had visited Beeckman in 1629 and like Descartes before him found inspiration in reading Beeckman's diary. He had started out as a sceptic with a penchant for attacking Aristotle's doctrines. Encouraged by his visit to Beeckman, he now saw it as his life's work to reconcile ancient atomism with Christian belief. Just as Descartes had done, he kept the book in which he recorded this pious undertaking to himself and dared to publish it only shortly before his death. Again the fear of censorship led to long-lasting self-censorship.

But this does not mean that there were no chances at all in France for ongoing innovation in nature-knowledge. Particularly in Paris during the conflict over Descartes' ideas, various informal groups of scholars met together regularly to discuss and exchange thoughts on the most wide-ranging phenomena and ideas. We have very few details since most kept their exchanges deliberately oral and unrecorded. But all kinds of things were happening informally between 1645 and 1660, and in Paris there was certainly a real ferment. However, the watchword was caution and visible outcomes were few.

In France, then, the picture is mixed. Elsewhere hardly even that. In Europe as a whole there was a widespread inclination to regard revolutionary nature-knowledge as bizarre to say the least, and more-over in conflict with fundamental values supported civilisation-wide. So in many cases the actual or potential bearers of strange and danger-ous nature-knowledge remained silent, be it voluntarily or from necessity. It had not quite reached the point yet, but it was not far off, when it would finally and definitively run out of steam. And the precedent of Islamic civilisation suggests that, although a later revival might certainly have been possible, it would be in a different spirit, as a reorientation towards a Golden Past rather than the exploration of as yet unknown, distant territory.

But just before it had reached the point where it ran out of steam, there was a turn-around. That turn was primarily due to the European great powers bringing to an end the conflict in Germany that had lasted thirty years from 1618 to 1648 and that had come perilously close to plunging Europe for ever into chaos and anarchy.

EUROPE'S NARROW ESCAPE

For more than a century almost every conflict that arose in Europe was quite quickly brought to a head. There seemed to be tinder in every nook and cranny. The causes of armed conflict were not simply such tangible ones as the Habsburg drive to European hegemony or the profits of overseas colonies. Time after time, philosophical and theological issues also led to conflicts or helped to aggravate them, with religion always at the centre. Christian faith is rich in dogmas on which opinions can easily differ, and since the Reformation those differences had been made deeper and more extreme. Typical is how, during the Twelve Year Truce between the Dutch and the Spanish, while Spain was still seen as the national enemy, an academic dispute about human free will in the face of Divine predestination very nearly escalated into a civil war. And we saw earlier in this chapter how closely religion and nature-knowledge had become intertwined. Galileo as well as Descartes had shaken the foundations of Aristotle's teachings, but these were closely bound up with Christian dogma. In an atmosphere in which the implications of every point of view could be taken to extremes, differences of opinion on natural phenomena, even on the question of why ice floats on water, had an innate chance of getting out of hand. Nothing was neutral in terms of one's broad conception of the world. Religion was obviously not, but neither was nature-knowledge. 'Athenian' natural philosophy was never neutral, and the 'Athenian-plus' philosophy of Descartes even less so. 'Alexandrian' mathematical nature-knowledge was dragged into the traditional territory of the natural philosophers by Galileo and Kepler and so ended up on the field of battle between conflicting world pictures. And that is how the future of both new forms of nature-knowledge came to depend on the outcome of that all-encompassing conflict that was being fought out in Europe.

In 1648 the European great powers concluded the Peace of Westphalia. It was a package deal which brought a whole series of concurrent conflicts to an end. Eighty years after the Dutch Revolt had broken out, the Dutch Republic was formally admitted to the 'concert of nations'. Germany remained divided into hundreds of princedoms and was left to its fate. The Austrian and Spanish Habsburgs declared themselves satisfied with what they had. Throughout Europe, every prince could henceforth decide on his territory's predominant religion without interference from outside.

The question was, of course, whether this continent-wide pact would hold. It did, and within about a decade the air cleared. In a certain sense the atmosphere in 1640 was more different from that of twenty years later in 1660 than from a century earlier in 1540. The cauldron had not exploded; the lid had been lifted just in time and the steam had been blown away. Of course, universal peace did not suddenly break out with the signing of the Peace of Westphalia. But a more measured response to conflict situations was adopted in a more conciliatory atmosphere, which made people more ready to look for an acceptable compromise. After a short delay, this also happened in England. The death of the Protestant dictator Oliver Cromwell and the rejection of his incompetent son laid the foundations for the return of the eldest son of the beheaded King Charles I. The restoration of the Stuarts, marked by the coronation of Charles II in 1660, went off remarkably smoothly. The new king was not particularly vengeful and slotted with relative ease into the new atmosphere of damping down opposing viewpoints and opinions.

The significance of all these events for the future of innovative pursuit of nature-knowledge was considerable. It could now become clear to outsiders, too, that these new modes of nature-knowledge did not just pose dangers but also opened up new opportunities. Those opportunities were of two kinds. Ideas could be developed and steps could be taken to isolate the philosophical-theological aspects from those that were, in that sense, neutral. It was precisely the latter that offered the chance of applying the new nature-knowledge to all kinds of useful ends. They included in particular waging war more effectively, putting the arts and crafts on a radically new footing, and increasing prosperity.

Where and how were these opportunities seen, assessed and actively adopted? We consider first the 'where' and then the 'how'.

Between 1660 and well into the eighteenth century the pursuit of the three new modes of nature-knowledge was mainly oriented towards the three cities of Rome, Paris and London. In a direct sense Rome's contribution was very limited, but it was here that the Jesuits had their headquarters. Wherever in the world Jesuit fathers occupied themselves with nature-knowledge they did so in accordance with centrally established and maintained guidelines. In contrast, Paris and London were cities where a great deal of research was carried out on the spot, much of which radiated across the rest of Europe. So in the second half of the seventeenth century the geographical centre of nature-knowledge was situated very differently from the previous two hundred years.

This shift from the Mediterranean to the Atlantic was part of a much wider development. Politically, commercially and culturally the European centre of gravity moved north. No longer was it the power claims of the Habsburgs that dictated European politics but those of the Bourbons. Control of the overseas trade routes became the object of fierce rivalry between England and France, the two great powers on the Atlantic coast. Those seeking prominence in literature, painting or music made sure to end up in Paris or London, and as close to the court as possible. In both courts a place was also secured for nature-knowledge. Indeed, more than just a place, for in both cities societies were formed specifically dedicated to the new style of natural research.

To make that possible, strategies had to be devised to strip away the more offensive aspects of these new modes of nature-knowledge. Some strategies were mutually exclusive, others complemented each other, but the objective was always the same, to make acceptable what had previously seemed unacceptable.

Atheism was not a real issue at this time; it existed in theory but in all probability it was a position held by few if any in practice. For any European who had not grown up in a Jewish quarter the norms for all his thought and behaviour were Christian. Voet's worries were therefore widely shared. There were many, Protestant and Catholic alike, who like him believed that at the end of the path taken by Galileo and Descartes might well lie the eventual downfall of Christendom. But was that a sufficient reason to reject their pioneering efforts? Even supporters of the new modes of nature-knowledge could sense the threat of atheism. Only, clinging with Voet to those well-worn grounds for proof of the Christian faith no longer appeared feasible in view of the new insights into the order of the natural world. Two possibilities remained. One was to return to revelation as the only reliable warrant of faith. That is what Pascal, a prominent practitioner of experimental-mathematical nature-knowledge, aimed for with his famous *Pensées*. But in that he remained practically alone. The other solution became enormously popular, lasted for centuries and now in its modern guise of 'Intelligent Design' has returned from never really having been away. Just look at how skilfully nature has been designed, how precisely all its parts, great or small, fit together and how carefully the laws of nature have been drafted! It cannot have just happened; it must have been done for us by a God who created us in His own image. In the seventeenth century dozens of treatises along those lines were published. Quite a few of them were written by practitioners of the new modes of nature-knowledge themselves.

At the headquarters of the Jesuits, who were so keen to take the lead in nature research but had chained themselves to Roman Catholic dogma in its entirety, the problem was approached differently. They first removed all the offensive aspects of the innovators' work and added the remainder to those portions of Aristotle's doctrine that still looked tenable. The resultant mishmash was then enriched with particles of matter, with experiments and with extensive calculations, and they imposed some kind of coherence on all this with ideas taken from natural magic. In their daily work they focused more and more on experimental research of the exploratory, fact-finding type. To explain the outcome of their experiments they could then fall back on their mishmash.

The Jesuits were not the only ones acute enough to perceive that experiment in itself is ideologically neutral. It was for that reason chiefly that most of the research carried out by the societies in Paris and London was largely experimental. At times, the experimenters even held back from attempting to explain the phenomena that they observed during their experiments. Explanation, after all, belonged to the old dogmatic natural philosophy that time after time had given rise to conflict.

The charters of both societies went a step further in the same direction. Both the members of the Académie Royale des Sciences and the Fellows of the Royal Society for the improvement of naturall knowledge by Experiment were specifically instructed by their monarchs to abstain from discussing politics, theology and philosophy, and to concentrate exclusively on natural research. The French Académie was more tightly organised and controlled and much more an instrument of state policy than its English counterpart. The young Louis XIV invested heavily in it; the members were a select group of leading researchers who were paid correspondingly high salaries. The term 'select' meant exactly that. Not only did the king want to bind leading talents to his academy, but within those leading talents he took care to make a further distinction. He was able to draw from the informal groups who had tried to keep the flame of natural research burning during the 1650s. But he excluded all those who used their talent to proclaim Descartes' natural philosophy to the world. Particles in motion were all very well, but not as part of a wholesale philosophy in the Athenian style. Jacques Rohault, a full-blooded

Cartesian, was a brilliant man and did the rounds of the Paris salons with great success. There he delivered elegant lectures in which, supported by spectacular experiments, he presented Descartes' ideas to the educated of, remarkably, both sexes. There were Parisians who, blessed with lesser gifts than he, were admitted to the weekly discussions of the members of the Académie, yet the court was implacable. Dogmatic natural philosophy only created discord, and Rohault remained firmly outside the gates of the Académie.

The consequences for natural research of the developments in 1660s Paris and London were immense. For the first time in history the study of the natural world became an autonomous activity undertaken by a more or less cohesive body of specialist scholars who were in daily contact with each other. In London, the degree of autonomy was greater than in Paris. There, the king could and did insist that his preferred research themes be addressed in exchange for the salaries that he paid the members, whereas Fellows of the Royal Society were largely free to pursue their research in any particular direction, whose outcome was neither preordained nor predictable. Furthermore, researchers could now release their results in a new type of publication: regularly published journals entirely devoted to the new style of natural research. By the same route, innovative researchers from outside France and England found an outlet and a breeding ground in the two great new centres of research. Christiaan Huygens became the nominal leader of the Académie as well as a corresponding Fellow of the Royal Society; Antonie van Leeuwenhoek recorded all of his microscopic discoveries in a series of letters to the Royal Society.

And so in the course of the 1660s the new spirit of the Peace of Westphalia succeeded in ridding the new modes of nature-knowledge of the all-pervasive fog of sacrilege. Even so, for a pious soul it remained an open question whether there were not all kinds of dangers still attached. But in the new decision-making centres in Europe, particularly the royal courts in the two greatest capital cities, this was no longer an issue. Natural research in its new guise had quite effectively been detached from religious and philosophical matters. But this still leaves wide open the question of *why* so much effort was put into it. Even if the latent danger of sacrilege in the new nature-knowledge had really been fully excised, could society not manage perfectly well without it?

For many in Europe, especially the still numerous Aristotelian or partly Aristotelian professors, the answer was obvious: yes please! But in many a princely court, particularly in London and Paris, a different opinion prevailed. Those in authority had been convinced that at least some aspects of the new nature-knowledge could be of material, practical use.

The idea that nature-knowledge could be practically useful was not at all new. In China it had never gone away; in Europe it arose in the mid fifteenth century with what I have called the 'third mode of nature-knowledge'. The research undertaken by Leonardo, Paracelsus and numerous others had centred on more or less presupposition-free observation directed where possible to practical usefulness. We have already seen that initially its practical uses were quite limited. Only in linear perspective, the building of fortresses and the determination of place on Earth had investigators managed to find interfaces between nature-knowledge (of the Alexandrian type) and craftsmanship. The Scientific Revolution of the seventeenth century is characterised by, at bottom, the same pattern, but on a larger scale, and in the long term a very different outcome. Great expectations were aroused of what nature-knowledge in its now drastically transformed guise might bring about for the arts and crafts and for warfare. Only, these expectations were largely disappointed, at least for the time being.

It had started with Galileo himself. Hardly had he discovered the four moons of Jupiter when he realised that their orbits could be used to solve the problem of longitude at sea. Unlike latitude, which is defined by distance from the equator, longitude is distance to a null-meridian (for which the Greenwich meridian has become standard). For any ship that was blown off course by a storm and might at any moment hit the rocks and go down with all hands, knowledge of its whereabouts would be invaluable. So it was not for nothing that royal sums were promised by the rulers of France, England, Spain and the Dutch Republic to anyone who could discover a means of establishing longitude with sufficient accuracy. Surely what, thanks to the Sun and the pole star, had been done for latitude could also be done for longitude? Galileo's hoped-for solution proved in the course of the century to be as unsatisfactory as two other theoretically sound solutions which in practice foundered on unforeseen problems which at the time proved insoluble. The second theoretical solution used the Moon's orbit while the third used the measurement of time. It is not by chance that the man who, in the generation following Galileo, seemed to hold the key to the two astronomical solutions, Giovanni Domenico Cassini, and Christiaan Huygens, the inventor of the pendulum clock, were precisely the two men whom Louis XIV managed to lure into joining his newly established Académie at the highest salaries then on offer. After all, the man who could solve the problem of longitude would enable his patron to control the seas.

In similar fashion Louis expected the mathematicians in his Académie to give the practice of gunnery a decisive edge. Again it had been Galileo who had immediately seen the practical significance of his discovery that the path of a cannonball is parabolic. In spite of the refinements that were added by his pupils and indeed his pupils' pupils in the following century, it achieved practical significance for warfare only in the time of Napoleon. And that is how it almost always went. Glauber, a disciple of van Helmont, tried to make fertiliser from wood juice. Robert Hooke, a prominent member of the Royal Society, thought that with the help of the new nature-knowledge he had found a means of making oil lamps burn more evenly. Across a broad front, not just warfare but also a wide range of crafts from machine construction to church organ building, experimental-mathematical and exploratory-experimental nature-knowledge seemed to hold everything required to restructure traditional craftsmanship from the ground up. Expectations remained high for a long time but, at least in the seventeenth century, they were but very rarely fulfilled.

Louis XIV and his minister of finance, Colbert, shared these expectations and invested a great deal in them. To be sure, this was not their sole motive for binding the cream of Europe's investigators to the French court. Their work would redound to the glory of the Bourbon dynasty in the same way as Molière's plays did, or Racine's poetry or Poussin's paintings or Lully's motets. What they expected specifically from the members of their Académie were practical benefits, and many of their commissions were directed to that end. Cassini was given the task of accurately remapping the whole of France. Louis had a sense of humour – when it turned out that earlier, less well-equipped and less well-trained surveyors had overestimated the size of his kingdom, he complained that his dearly paid astronomers cost him more land than his generals had managed to conquer for him.

Louis further had some patience, in the same way that we have nowadays with research teams who having spent their initial subsidy claim to be on the brink of making a breakthrough well worth the next down payment. But that patience cannot sufficiently explain the huge sums spent on the Académie which, viewed with a sober mind, failed to deliver a great deal, and certainly delivered a great deal less than promised.

The point is that it was not really 'viewed with a sober mind'. The widespread myth of the enormous boost that the rise of modern science immediately gave to Europe's material prosperity is a myth that dates from the time itself. Much historical research into the matter has in its turn been deeply affected by that myth. What interests us is its origin and how it generated so much support in the seventeenth century. Or rather, it was not so much a myth as an ideology. By 'ideology' I mean here a compound of views on tangible reality which at the same time rises above reality in an all-encompassing vision. We shall name it after the man whose name is most closely associated with this vision and call it the Baconian Ideology.

Let me first summarise the line of thought that has brought us to this point. The new modes of nature-knowledge managed against many odds to survive the crisis of legitimacy that threatened to overwhelm them, thanks to the neutralisation of the theological and philosophical dangers that affected them, and further to their hoped-for material usefulness. The latter is what made the effort put into the former worthwhile. The circumstance that hardly any use really came from the many efforts undertaken to that end was clearly a problem, or rather it would have been a problem if it had been noticed to any substantial degree. That it was not, and that high expectations continued into the eighteenth century when they eventually started to be met, must be attributed to the rise of the Baconian Ideology and to the wide support it received.

That ideology can be characterised as a leap of faith in the power of the new nature-knowledge. 'Knowledge is power', as Bacon had so eloquently argued. All of his work was imbued with a sturdy conviction that, in its new guise, nature-knowledge was capable of 'the effecting of all things possible'. His writing started to strike chords during the English Civil War, and all kinds of half-baked utopian programmes for improving the world were based upon them. In the 1660s, after the Stuart Restoration, this at times intemperate body of ideas was reined in to provide theoretical justification for the efforts made by the Fellows of the Royal Society. 'Justification' here naturally means in the first place demonstrating that their work was in accordance with Christian doctrine. In particular, the objection that all that experimentation merely distracted people from directing their immortal souls to the hereafter had to be refuted. Two Anglican priests took that task upon themselves. The future Bishop Sprat argued in a characteristic passage that there are not one but many ways of serving the Lord. Those whose character does not lend itself to turning their backs on the world may compensate for it through experimental enquiry that will enable them to be serviceable to the world. Was not Jesus, who used to withdraw from the company of others when bent on resolving an inner conflict, in the habit of using for his conversions 'some visible good Work, in the sight of the Multitude'? Is not the systematic correction of errors and mistakes in experimental research the secular counterpart to the ascetic's spiritual penance?

> Seeing the Law of Reason intends the happiness and security of
> mankind in this life; and the Christian religion pursues the same
> ends, both in this and a future life; they are so far from being
> opposite one to another, that Religion may properly be styl'd
> the best and the noblest part, the perfection and the crown of the
> Law of Nature.

In short, Christianity in its full extent and nature-knowledge in its
new form complement each other in a natural manner.

Here and elsewhere in Sprat's argument we encounter something world-historically unique. At the end of the previous chapter
I argued in the footsteps of Max Weber that, in contrast to the
other religions of the world, religious experience in Europe was being
directed less and less to a mystic turning inwards. In the west
European variant of Christianity, and particularly among Protestants,
abstaining from worldly pleasures for the sake of a place in the hereafter increasingly took the form of frugality, diligent industry and a
sober sense of enterprise. Whether this actually did have the effects on
the spirit of capitalism which Weber sketched a century ago I shall put
to one side. But it is obvious that this is exactly the tenor of Sprat's
apology for the Royal Society. He linked that attitude especially to the
new exploratory forms of experimental nature-knowledge. And with
that there emerged a *religious* sanctioning of sheer *worldly* knowledge
which had never been seen anywhere. In particular, nothing like that
had ever been seen in Islamic civilisation.

Surely, in Baghdad under the Abbasids a relatively open,
outward-looking form of Islam set the tone, in which Greek nature-
knowledge could be adopted and have new, exciting life breathed into
it. A sourpuss and complainer like Ibn Qutayba, who thought that the
only knowledge worth having was in the Koran and that 'foreign'
nature-knowledge was good for nothing anyway, easily lost out in
the general enthusiasm. But hardly had the wave of invasions led to
a turning inwards than the Ibn Qutaybas of this world resurfaced and
the whole enterprise came to an end. That it could happen so quickly

was at least in part because the practitioners of 'foreign' nature-knowledge did not have the means at their disposal for creating an ideology to fall back on in dire times. In a comparable crisis of legitimacy, seventeenth-century Europe did have such means available and did create such an ideology. It was especially the case in Britain which was emerging more and more as the main powerhouse of European nature-knowledge. The Baconian Ideology took root there; upwardly mobile social groups involved in navigation and trade embraced it as a symbol of progress. The Baconian Ideology remained a British product. On the continent, in the Académie for instance, lip service was occasionally paid. But it never really caught on, not even in the Northern Netherlands, which is the more remarkable in view of the strongly Protestant component of the Baconian Ideology. As a Dutchman I harbour a secret suspicion that in the Netherlands usefulness was already being understood as immediate gain, whereas at the core of the Baconian Ideology lay the confident expectation of gains in the long run, not necessarily tomorrow. Knowledge, to use modern jargon, is certainly capable of being 'valorised', but very seldom in the immediate future and certainly not in predictable directions or on demand.

It is time to come to a conclusion, indeed to three conclusions.

In the first place, we can now provide a fairly precise answer to the frequently asked question of what the Reformation meant for the Scientific Revolution. It is true that rather more Catholics than Protestants contributed to the revolutionary transformation of the original Greek modes of nature-knowledge and their later expansion, but the numbers roughly reflect their relative share of the population of Europe as a whole. The third, exploratory-experimental mode of nature-knowledge ended up in the hands of Jesuits, and in Paris mainly of Catholics while mainly of Protestants in London. It was most fruitfully applied, as we shall see, in Britain under the banner of the Baconian Ideology. But above all, the firm belief in the productive potential of the new modes of nature-knowledge and their latent power radically to transform human destiny by providing wealth and mastery over nature was a Protestant affair.

In the second place, we can now see that even if Greek nature-knowledge had produced a Galileo-like figure in the civilisation of Islam, such a figure would not have been able to maintain the pace and keep the flame burning. The philosophical-theological consequences attached to such an appearance were insurmountable. They would have encountered comparable resistance in any civilisation with a Holy Book. Only, Islam was not in possession of the kind of full-scale theology of possibly non-literal exegesis which St Augustine had introduced into Christendom. Nor did Islam have the extroverted means of salvation to hand on which an ideology like the Baconian could be based. Christian Europe was unique in its strongly outward orientation. There is no special merit in this; it is just a sober, well-observable historical given. So to the frequently asked question of why there was no Scientific Revolution in Islamic civilisation, the answer is twofold. Without the wave of invasions an initial move in the direction of an 'Alexandria-plus' of a kind might well have taken place – some effort of the historical imagination suffices to make an 'al-Galileo' appear in the mind's eye. But Islam did not have at its disposal such resources as would have been necessarily required to create an ideology fit to cope with the otherwise lethal philosophical and theological consequences of his ideas. In the spectrum of inward-looking to outwardly oriented, Islam was and is a religion of the centre – neither so extremely introverted as Hinduism in India, nor so uniquely extroverted as the direction that Christianity had taken more and more in Europe. An 'al-Bacon', with the best will in the world, cannot even be imagined.

Finally, and this is our third conclusion, between about 1645 and 1660 even extrovert Europe only just slipped through by the skin of its teeth. The Peace of Westphalia and the shift of the European centre of gravity from the Mediterranean to the Atlantic created a framework in which the crisis of legitimacy that hit the new modes of nature-knowledge could be resolved. The rulers who were decisive when Europe's fate hung in the balance grabbed the opportunity that they created for themselves on the brink of total chaos.

They also saw the potential uses of the new nature-knowledge and deliberately helped to isolate its sacrilegious aspects and render them harmless. On top of all this an ideology came up which, despite the failure of the new nature-knowledge to be of any immediate use, prevented it from coming to a complete halt. All in all, this amounts to quite a bit that in circumstances that were no more than slightly different might well have turned out quite differently. We saw earlier that the three revolutionary transformations around 1600 were far from being a historical necessity. We have now seen that there was even more coincidence in the survival of the three new modes of nature-knowledge than in their initial appearance.

In the present chapter we have discussed the survival of those new modes from the point of view of their anchoring in society. European civilisation, like every other, ran on certain fundamental values, and it had proved possible to make it plausible on a sufficiently wide scale that the new modes of nature-knowledge fitted in with those values, or even contributed to their expression.

But that civilisation-wide value correspondence does not provide the only basis on which the new modes could survive. As we shall now see, each was also marked by a dynamic of its own; each also made advances that to some degree were autonomous and driven from within.

5   Expansion, threefold

Kepler and Galileo, Beeckman and Descartes, Bacon, Gilbert, Harvey and van Helmont were each in their own way pioneers who, with their revolutionary transformations around 1600, were far in advance of the foot soldiers they had left behind. Even if we leave out of further consideration the profound ideological problems attached to their work, one would still have expected the three prevailing modes of nature-knowledge to survive for some time as if nothing much had happened in the intervening period. The surprising thing is that when you look back from around 1700 you find how little remained of these 'old' modes of nature-knowledge. In the first half of the seventeenth century there were still many who practised mathematical nature-knowledge in the old, hyper-abstract manner of Archimedes, of Euclid and of Ptolemy's *Almagest*. This was particularly true of planetary theory and the study of light and vision. When, between 1600 and 1625, Harriot, Snel and Descartes each discovered the law of refraction first found by Ibn Sahl, it still fitted almost seamlessly within the tradition. And in so far as a generation of astronomers after Kepler could approve of his results at all, they were as a rule still handled in tried and trusted fashion as fictional aids to model building of the usual kind. By the end of the century, this has all been swept away and 'Alexandria-plus' is the order of the day. This involves what I have called Ptolemy's bridge building as well. Ptolemy had ingeniously sought to inject elements of the real world in current, highly abstract mathematical accounts of the propagation of light (updated in the eleventh century by Ibn al-Haytham), of musical consonance (updated in the sixteenth century by Zarlino) and of planetary trajectories. As realist-mathematical science expands beyond Kepler and Galileo, all this begins to be replaced with almost brand-new physical theories in the former two cases, concurrent with the speedy demise (at least among astronomers) of belief in horoscopes or in any planetary influences at all. When in December 1659 Huygens is invited by a daughter of the late *stadhouder* (whom his father had served as secretary) to inspect her nativity and cast her horoscope, he knows that he cannot refuse, but he asks his fellow astronomer Boulliau to do it for him. One generation later, by the end of the century, even courteous gestures like this are on their way out (although to append a fixed date to when astrology ceased for good to be of intellectual interest to astronomers is still being debated by specialist historians).

The victory of 'Athens-plus', in partial contrast, is not quite so comprehensive, at least not in the universities or among the Jesuits. There, various mixed forms make their appearance while pure Aristotelian doctrine becomes involved in all kinds of rearguard actions. Finally, exploratory, fact-finding experiment has never completely replaced straightforward observation, but that has more to do with the nature of what is being observed – by the end of the seventeenth century there is no hesitation in performing, or at least supporting, an observation with experimental means whenever the opportunity presents itself. And even the barriers that separate the different modes of nature-knowledge start to crumble. While each mode continues to flourish independently of the others, they also start to combine in productive ways.

In the Old World, where innovation was not as routine as it is now but still very much the exception, we can only call this a lightning-fast victory of the new over the old. Large-scale princely support attained in the 1660s, and the Baconian Ideology which, especially in Britain, gave wings to the effort, certainly contributed a great deal. But what *intrinsic, substantive* developments furthered the victory? We shall consider each of the transformed modes of nature-knowledge in turn.

'ALEXANDRIA-PLUS' SPREADS LIKE WILDFIRE

We have already noted that the transformation of 'Alexandria' into 'Alexandria-plus', brought about by Kepler and Galileo, introduced greater depth as well as a considerable widening of the mathematical mode of nature-knowledge. After their death, the process was carried forward enthusiastically by but a very small number of successors. To give an idea of how this worked in practice I shall take two examples. One, taken from nature, concerns the void; the other, taken from craft practice, is about the taming of rivers. Both topics had been investigated or at least thought about previously. The examples illustrate not only what a mathematical approach could contribute to their solution, but also the limitations that the approach was as yet subject to.

Galileo's *Discorsi* opens with a recollection of the time when he was professor in Padua and regularly visited the workshops of the Venetian fleet (the Arsenal). There he noticed that the workmen could not manage to pump water higher than about 10 metres. Anything more and the column of water 'broke'. They themselves assumed that there must be something wrong with the material of which the pump was made, and accepted it as one of those facts of life that one just has to live with and make the best of.

Galileo instead sought to link the phenomenon to nature-knowledge. He attributed it to something that he called 'the resistance of the vacuum'. Shortly after his death in 1642 his pupil Evangelista Torricelli continued the work. He conjectured that the column of water in the pump is counterbalanced by the column of atmospheric air pressing down on the water. Equilibrium is reached when the water column is about 10 metres high. But if that really were the case, he reasoned, you could make the phenomenon much more manageable by using a heavier liquid. And indeed, if you fill a test tube with mercury and hold it upside-down in a dish of mercury, the column of mercury only reaches about 76 cm above the level of mercury in the dish. In the space above the mercury in the tube, there is nothing; it is just empty.

That space was later named after Torricelli. Unfortunately, with this experiment he moved on to very thin ice, not only in terms of natural philosophy but also theologically. Had not Aristotle given conclusive proof that empty space not merely does not, but cannot, exist? Furthermore, once a void is mentioned, atomism is not far behind. In an Italy under the ideological thumb of the Inquisition, it would be safer for Torricelli to hold his tongue. Which is what he did.

The research was taken up in France by a young scholar named Blaise Pascal, an eminent adherent of experimental-mathematical nature-knowledge. He was also a fierce opponent of the Jesuits and their theology, which appeared to place rational evidence above divine revelation. To him the question of whether Torricelli's space is empty or not offered the perfect opportunity to demonstrate the superiority of the new mode of nature-knowledge over the sterile natural philosophy on which his enemies relied. He assembled his experimental and reasoned evidence for the void, step by careful step, up to and including a final test, which involved climbing a mountain. Blaise's sister lived with her husband at the foot of the Puy-de-Dôme in central France. At the request of his Parisian brother-in-law, Florin Périer filled two tubes with mercury, placed them upside-down in a tray of mercury, left one of them behind at home and climbed the mountain with the other in his hand. Pascal had reasoned that if the columns of mercury and air are in equilibrium, then to reduce the column of air by climbing a mountain should make the level of mercury drop. And that is exactly what Périer was able to confirm in painstaking detail to his brother-in-law.

His conclusion that Torricellian space is empty and that the effect is due to air pressure led him into a polemic with the Jesuit Father Étienne Noël. Another scholar whose approach to natural phenomena Pascal thoroughly disliked had once been a pupil of the elderly Father. His name was René Descartes. The basic principles of his natural philosophy excluded the possibility of a vacuum quite as firmly as Aristotle's had, although for different reasons – as we know, according to Descartes space and matter are identical. And so Pascal was able to kill three birds with one stone: Aristotle, Descartes and the Jesuits who created such an odd concoction of the two natural philosophies. In a more general vein Pascal took the opportunity to

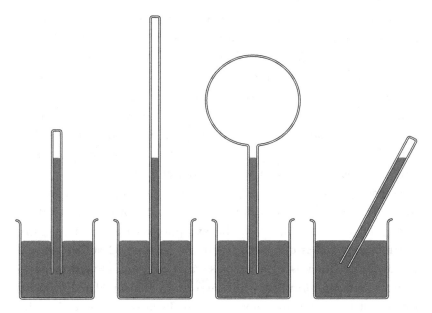

Schematic representation of Torricelli's experiment
Far left, Torricelli's experiment. The remaining three images show
Pascal's experiments demonstrating that the height of the column of
mercury is not affected by the length, shape or angle of the tube.

point out the shortcomings of speculative natural philosophers and
their know-all pretensions in comparison with the modestly testing,
experimental researcher who proceeds step by step and does not claim
to know beforehand how everything fits together. Even today, Pascal's
letter to Noël is a compelling piece of prose. Their brilliant rhetoric
and occasional exaggeration notwithstanding, those seven pages
written in crystal-clear French and full of memorable one-liners form
a glittering manifesto of modern natural science in the making.

Through the Po delta flows a river, the Reno. Past Bologna, in
the neighbourhood of Ferrara, it comes out in an arm of the Po, but in
the seventeenth century it had been diverted into a marshland area.
The consequences had been unexpected and catastrophic, leading to
the loss of many acres of farmland. How and where could the Reno be
re-routed back into the Po? In 1625 the authorities of the papal state of

Rome, of which Ferrara and Bologna were then a part, invited Benedetto Castelli, a monk and student of Galileo's, to find a solution. Castelli responded in the style of his teacher. He formulated a general law which established a connection between three variables as follows: 'The cross-sections of the same River discharge equal quantities of water in equal times, even though the cross-sections themselves are unequal.' This law and its basic assumptions resulted in very different recommendations from those made by the Jesuit advisers. They preferred to approach the problem not with a single sweeping generalisation, but by empirically investigating the various factors that might affect the flow of the water, followed by a quantitative evaluation. Remarkably, in this exceptional case these two vastly different approaches did come together over time. Castelli's successors included other aspects, such as water pressure, in their geometrically expressed formulas, while later Jesuit advisers dared to make somewhat more forceful generalisations. Even so, at the end of the century yet another committee of cardinals which was convened to tackle the problem was still at a loss what to do. The decision-making process was really dominated more by the conflicting interests of Bologna and Ferrara than by the question of whether the mathematical model of one party or the reasoned estimates of the other best reflected the Reno situation. Even in our day, a decision to adopt some mathematical model that has been devised to help solve a practical problem is not made automatically, nor should it be. But it is also true that how we have learned to think about these things has meanwhile become so much more refined that to act solely on the basis of practical experience is no longer a viable option.

It was not only the vacuum and the diversion of river courses that were taken over by the experimental-mathematical approach. The change from 'Alexandria' into 'Alexandria-plus' opened a whole range of floodgates. Galileo himself was a case in point. He went well beyond falling objects, projectiles and a new account of motion. The number of topics that he was the first to broach mathematically grew to about a dozen. Even more important were the additional tools that he employed. Previously, the mathematical treatment of a problem was conducted in just one fashion, by means of abstraction via Euclidean geometry. In Galileo's hands this limited repertoire was expanded to include four quite new pathways towards the mathematisation of phenomena.

One of them is analogy. To gain an understanding of motion of one particular kind, you can take an already familiar kind as a starting point and find out how far you get by applying the mathematical rules that are valid for the one now also to the other. This is how from a consideration of equilibrium situations as his point of departure Galileo was ultimately able to arrive at his insight that uniform motion tends to be retained indefinitely. In similar fashion he tried, though unsuccessfully, to come to grips through mathematics with the blow of a hammer driving a post into the ground by comparing it with the pressure of a weight placed upon that post.

Then there is the reduction of a complex phenomenon to a mathematical model that is claimed to encapsulate the core of it. That form of mathematisation was adopted by Brother Castelli and his successors in the case of the Reno. With Galileo, and even more imaginatively with Kepler, we further see how in their handling of aspects of the real world they begin to work, and to calculate, with the infinitesimally small. Within two generations such efforts culminated in the discovery of the calculus by Newton and Leibniz. And finally, there is, of course, mathematisation through experimental testing. Analogy, mathematical modelling, the mathematics of the infinite, and experiment were all enormously powerful aids to gaining a grasp of the real world and are still very much in use by modern scientists.

In the generation after Galileo all this was checked over, corrected, elaborated, refined, subjected to new mathematical techniques and also extended to other fields of mathematical research. Moreover, the five traditional Alexandrian subjects were now taken up, one by one, within the ambit of the new abstract-realist approach, as for instance with consonant intervals being treated henceforth as determined by vibrational frequency. The same applies to a range of medieval initiatives which creative Aristotelians such as Buridan and Oresme had introduced. Prised loose from the framework of natural philosophy and absorbed into the new framework of 'Alexandria-plus', vague notions such as 'impetus' could now be defined with sufficient exactitude to help draw up new rules of motion.

The breadth and depth of these developments are all the more astonishing when we consider how few people after Galileo were actually involved. In Italy, due to the limitations imposed by the Inquisition, there were only Torricelli and a handful of students whom Galileo had gathered around him in Florence. A little later in France, about the same number of members of the Académie had a penchant for mathematics, Huygens among them. Here and there a few loners made important contributions, like Pascal in France and Horrocks in England. Perforce, the Jesuits remained outside – one of the most serious losses brought about by the controversy over the motion of the Earth. Had it not been for that controversy, Clavius' intellectual heirs might very well have made the transition from natural philosophy cultivated the quantitative way to a fully mathematical treatment of real phenomena within one or maybe two generations. In the prevailing circumstances they were obliged to step back and cling to the approach that we saw them adopt with the river Reno: cautious generalisation based on empirical data accepted in a critical or not so critical vein and enriched at times by measurements.

The flow of events surrounding the continually postponed diversion of the river Reno is but one of many cases in which little or nothing came of the claim that the new realist-mathematical nature-know-ledge would place the arts and crafts and their tried and trusted methods on a radically new footing. If we map out the dozen or so areas where a serious effort was made to realise those pretensions in the seventeenth century we see in the first place that the urgency of a problem made no difference whatsoever to success or failure in finding a mathematical solution. Diverting rivers, improving gunnery and establishing longitude at sea were pressing problems. Lives and prosperity were at stake, and the authorities were prepared to invest a great deal of time and money into resolving them, yet unsolved they remained. The longitude problem, for instance, would have to wait until halfway through the eighteenth century before John Harrison overcame the practical obstacles with his carefully constructed, seaworthy timepieces.

Why was this so? The point is that the gulf between crafts and mathematical nature-knowledge was still far too wide to bridge at the first attempt. On the mathematical side of the gulf, the models that were devised took as yet insufficient account of all the factors relevant to a particular situation, as was so clearly the case with the Reno. The world is far too untidy for a single model to be adequate. As a rule, behind every natural event a multiplicity of diverse patterns are at work. Very often the mathematical techniques were not yet up to the task at hand. On that point, Leibniz's discovery and subsequent handling of the differential and integral calculus towards the end of the century improved matters greatly. On the other side of the divide, craftsmen were at times quite right to ignore the prescriptions handed to them by a mathematical scientist. For instance, a particular tuning system which the organist Andreas Werckmeister designed lacked any mathematical elegance but in practice it worked far better than the elegant solution to the tuning problem which Huygens came up with. On other occasions craftsmen clung too stubbornly to their trusted methods. The same Huygens complained, in connection with the trials for his sea-going timepieces, that the captains overseeing the tests had to put up with 'the crew's frequent scolding and mockery of this effort to measure longitude in a new way'. Craftsmen further tended to lack the required minimum of mathematical knowledge. And finally social distance was almost always too great for truly fruitful exchanges.

Communication was most effective when the mathematician's self-interest, the material tools of his own trade, were at stake. Along with the telescope, another mathematical instrument invented in the seventeenth century was the pendulum clock. Once again the idea came from Galileo who, as we have seen earlier, noticed that a pendulum of a given length swings to and fro in the same time period however great or small the amplitude (width of the pendulum's swing). In other words, it does not matter how far from the vertical the pendulum is released, it will always take about the same time to return to its starting point. But it was Huygens, not Galileo, who succeeded in coupling a pendulum to the mechanical clock in such a way as to make feasible a timepiece of unprecedented accuracy. In 1657 under Huygens' instructions, Salomon Coster, a clockmaker in the Hague, built the first pendulum clock. A rare copy adorns the walls of the Museum Boerhaave in nearby Leiden. In daily use the pendulum clock meant a vast gain: accuracy shot up from within about 15 minutes a day to less than 10 seconds. Astronomy also benefited greatly from more accurate time measurement. And Huygens' invention had even more consequences. The pendulum clock seemed to open the way to solving the longitude problem – and indeed a century later in Harrison's hands it did just that. Furthermore, the pendulum clock made Huygens realise that strictly speaking the pendulum's period (the duration of one swing to and fro) is truly independent of its amplitude only when the pendulum is forced to follow a special kind of path, that of a cycloid (the curve traced by a point on the rim of a circle as it rolls along a straight line). That discovery, in his own opinion the most important that he ever made, led him into a previously unexplored area of mathematics. To find proof for his discovery he was compelled to devise new mathematical techniques that, apart from the crucial lack of a general formulation, were effectively those of the later calculus. He also formulated what we still know as the law of the pendulum, the formula that binds together all the factors on which the period of a pendulum depends.

The pendulum is just one illustration of the subterranean connections between different fields which mathematical treatment by and after Galileo revealed. In that interdependence rested a tremendous dynamic. Advance in one field could easily lead a gifted researcher to rapid advance in another, apparently unrelated field. Huygens discovered and more or less solved most of the problems that he came across in connection with the pendulum in a creative burst of less than three months. The problem that he faced on 21 October 1659 still looked entirely unrelated to the central discovery of which he proudly informed his old teacher on 6 December. As so often happens in the history of creative thought, he found something he had not been looking for.

In principle such a sequence from problem to discovery to new problem to new discovery need never come to an end. It was certainly becoming clear that the realist-mathematical mode of nature-knowledge is of the 'open-ended' type. Even when a particular line of research seems to have been concluded, new perspectives have a habit of opening up. Neither does it have to be limited to the head and hands of one individual researcher. More and more, the publication of a particular discovery encouraged scholars elsewhere to pursue it further. The presence of two research centres in Paris and London provided an enormous stimulus (Paris was rather better for mathematical nature-knowledge than London). The official weekly sessions had already given rise to lively exchanges of ideas. A leisurely walk through the Jardin des Plantes or a lively discussion in a coffee house round the corner from the Royal Society could spark off new ideas. Monthly journals delivered on horseback and an effective postal service ensured that the exchange of ideas was not confined to Paris and London but that researchers in every corner of Europe could take part with ease. I argued earlier that the transformation from 'Alexandria' to 'Alexandria-plus' might have been possible in a manuscript culture – a statement by and large valid for the other two modes as well. But the expansion which took place in the second half of the seventeenth century and the type of to-and-fro exchange which gave it added vigour are wholly unthinkable without the printing press.

A further two dynamic elements contributed to the unprecedentedly rapid advance of realist-mathematical nature-knowledge.

One was that it proved adaptable and could even be restructured. Kepler's three laws were hardly presented to the public on a plate. Two were in *New Astronomy*, one in *Harmony of the World*. All three were almost buried under a mass of other topics, some of which looked quite dubious, especially Kepler's conception of forces in the solar system and his conclusions on how the harmonic intervals were expressed in the planetary orbits. Even in his later textbook those 'laws' hardly stand out. It was a child prodigy named Jeremiah Horrocks who managed to prise them loose a few years after Kepler's death, so that what became known as Kepler's laws we owe in some measure to Horrocks. Nothing like this had ever happened before in nature-knowledge. The nearest had been the occasional blending of components from two or more disparate systems of natural philosophy (a late example being the Jesuit mishmash), but the results lacked all coherence. In Kepler's case, by contrast, the fertile sections of his complex corpus of ideas were extracted, thus enabling them to come to full fruition. That is how it still works in modern science – the formula may remain the same but the meaning attached to it can change drastically. *That is, so long as the formula is correct.*

And with that we come to the last and most dynamic element in the process of widening and deepening that went on and on in the new realist-mathematical mode of nature-knowledge. This dynamic lurks in the complex relationship between a proposed mathematical regularity and its reality check. Galileo addressed the relationship twice in so many words.

In the *Discorsi* we find:

> The knowledge of one single effect acquired through its causes opens the mind to the understanding and certainty of other effects without need of recourse to experiments.

Elsewhere in the *Discorsi* he says:

> To deal with such matters scientifically it is necessary to abstract
> from them. We must find and demonstrate conclusions abstracted
> from the impediments, in order to make use of them in practice
> *under those limitations that experience will teach us* [italics added].

In the first passage, then, he asserts that once you have nailed down
the mathematical rule in one particular instance, you can apply it
elsewhere without any further experimental testing since you have
now got hold of the truth of the matter. In the other he states, much
more circumspectly, that experience will teach us to what limitations
the regularity now found is subject in the real world of everyday
practice. The first passage is closer to Galileo's fundamental conviction that the nature of reality is ultimately mathematical and also
reflects more closely his own actual research preferences.

But that is not the main point. The main point is that in these
two passages Galileo has marked out with striking precision the field
of tension in which, then as now, natural scientists have to operate,
especially those with a mathematical orientation. The outcome of an
experiment may mislead you by throwing up an annoyingly irrelevant
detail that obscures the underlying mathematical pattern. Mathematical deductions may take you down the wrong path due to the circumstance that so much in the natural world is unpredictably untidy.
Sometimes you are well advised to regard the result of an experiment
as a rebuttal of one theory in so far as it points the way towards a
better one. Sometimes you are well advised to ignore a negative result
at least temporarily. Either way, the human mind is always able to
find a way round a conclusion, however compelling. Falsification, in
the sense of irrevocably disproving a statement, never clinches the
matter forever. But that is not what is really at stake here. What a
handful of mathematical natural scientists began to discover in the
seventeenth century was the field of tension itself, that unceasing
interaction between mathematical order and attempts at experimental validation. It soon became apparent that there are no hard-and-fast

rules for falsification, only a firm conviction that the interaction can take us further. The precise balance between a general regularity and its experimental validation has to be fought out afresh on every single occasion, for each time it will prove to lie at a different point between the two poles that Galileo marked so presciently in the *Discorsi*.

Meanwhile, all this had introduced something fundamentally new into human thinking: systematic feedback. We present nature with a direct, precise question, and *nature gives an answer*. That answer can be very exact and unambiguous, as with the deviation of 8 minutes which persuaded Kepler to discard his own cherished hypothesis. The answer can also be, as I have just argued, more sphinx-like, leaving us wondering for quite a while what to do with it. To be sure, you may keep ignoring it – as human beings we can persuade ourselves of anything and keep on deceiving ourselves. We are very good at maintaining things which are manifestly not so, particularly when they involve us personally (Galileo's account of the tides is a somewhat painful example). The most extraordinary feature of the procedures developed during the Scientific Revolution in the realist-mathematical mode of nature-knowledge is that they went even further than the intended critical test. They gave room to such checks by filtering out the personal factor to the largest possible extent. What one person may maintain, and keep maintaining through stubbornness or pride or self-interest, another can cool-headedly retrace and, if the outcome gives occasion for it, find wanting.

Apart from 'Alexandrian' mathematical science, which was little concerned with reality, it had always been possible in every existing mode of nature-knowledge to assert something and cling to that assertion for so long as it sounded the least bit plausible. Nothing could be disproved conclusively because there was almost always something that could be said in favour of an assertion as well as against it. The outcome is then stalemate and the argument can last for centuries. One example is empty space, which the followers of both Aristotle and Descartes knew could not exist but which was utterly basic to the atomists. It was all based on reasoning, with a few relevant-looking facts thrown in for support. Pascal broke with all that and produced a well-designed experiment which, with the help of his brother-in-law, compelled nature to come up with a very specific answer. Of course, a natural philosopher could wriggle out of that answer if he so wished. Experiments of the validating kind cannot be conclusive until a human being decides that they are conclusive. But what they do accomplish is to raise the debate about right or wrong to an altogether higher plane.

In the world of the craftsman with its hard, material basis it had always been different. True, there is usually more than one apparently appropriate solution for a particular technical problem, but that some solutions just do not work shows up quite quickly in hard, everyday practice. If an organ pipe squeaks or a bridge collapses, you know that you have to do something differently. This kind of unforgiving feed-back had never previously been known in nature-knowledge, and in the realist-mathematical nature-knowledge of the seventeenth century its possibilities and limitations were now being explored. For the first time in history an opportunity had opened up to acquire knowledge which is not just plausible but which, whether or not it is correct in every single case, at least holds water.

'ATHENS-PLUS' CATCHES ON

Natural philosophy too showed a certain dynamism in the course of the seventeenth century, even though it took on rather different forms. The philosophy of particles in motion as developed by Descartes did not just give rise to fear and aversion. For many scholars it exercised a tremendous power of attraction. In *both* camps, for and against, there were laymen as well as vicars and even, here and there, a priest or a monk. As with the realist-mathematical mode of nature-knowledge there is no question of opposition only from the clergy and support only from laymen, or even atheists (of whom in fact there were none). What was it in the natural philosophy of Descartes that could move people to agreement and even enthusiasm? How did it overcome all its natural philosophical rivals with such remarkable ease?

We must first of all remember that victory over other natural philosophies was in itself nothing new. At one point the Stoa dominated, then Platonism, then Aristotelianism and in the second half of the seventeenth century it was the turn of atomism for a change, even though it came with the special 'plus' twist that Descartes had given it. But just as all those earlier changes of the guard had had their own particular causes, so we may reasonably ask what specific qualities made this new doctrine so attractive.

In contrast to the other natural philosophies, atomism had from time immemorial painted a world quite different from the way our senses perceive it. This was even truer of Descartes' doctrine of particles in motion. What we perceive as a tree would be a collection of particles of matter lumped together. What we observe as the burning of that tree is not (as with Aristotle) the visible transformation of earthly and watery matter into airy smoke and fire. It is an event that as such cannot be observed, in which the diverse particles of matter are torn apart through certain specific motions. According to the doctrine of moving particles, reality has a deep structure of which our senses can only provide indirect information. In no way are they able to plumb the depths of that reality.

This view of the world was greatly supported by the circumstance that scholarly Europe was learning precisely the same lesson at exactly the same time from a completely different source. Two new instruments, the telescope and the microscope, each in their own way, were showing up the limitations of our senses. From the stars that were seen to form the Milky Way to the delicate structures which even a moderately powerful microscope revealed, it was clear that our human powers of observation are inadequate without help. So was Descartes right after all? It seemed so. For several decades the early users of the microscope kept hoping and expecting that, as its resolving power kept increasing, the particles of matter which Descartes had postulated in so visionary a manner would eventually become visible. And as for the discoveries that Galileo and others had made with the telescope, Descartes himself had skilfully worked them into the picture of an infinite universe that he had sketched in *Le Monde* and later in his *Principia philosophiae*.

With the mathematicians the debate about the Earth's motion had been rather technical. Unless you had fairly precise knowledge about comets or the phases of Venus and all those other telescopic discoveries the debate was unlikely to make much sense to you. Descartes, in contrast, had made all that and a great deal more part of a picture of the world that anyone with an academic education could easily understand. More than that, to explain any particular phenomenon everyone could design their own mechanism. You only had to dream up a specific type of particle in a specific configuration moving in a specific way. If you wanted to contribute to the renewal of nature-knowledge you were not necessarily tied to the esoteric language of mathematics. Descartes' philosophy offered a parlour game from which no one need feel excluded. A stream of academic textbooks in the second half of the seventeenth century made grateful use of the opportunity.

Another attractive feature of Descartes' ideas was the ambiguous nature of his innovation. Certainly, he had called eloquently for systematic doubt and thereby offered to human thinking a degree of extra leeway which shocked Voet but gave many others a sense of liberation. Galileo had in his own way done much the same thing in his *Dialogo*. For him, however, the outcome of the new ways of research that he advocated lay in a distant, unknown future. For many people that was just a bridge too far. Having to think for yourself entirely independently may sound grand but when all solid ground is swept from under your feet, where does it leave you? What made Descartes attractive was that his challenging invitation to start thinking for yourself came accompanied by the comforting reassurance 'I have already done it for you.' What was so nice about Descartes was that he gave you the opportunity to wander through the forest on your own, with no authority holding your hand to guide you down well-trodden paths, and yet as soon as you risked losing your way a roadmap popped up, ready to show you in which direction to proceed. Safe innovation: what person not completely wedded to tradition or to the teachings of his church or to both would not succumb? In the seventeenth century there were few who did not.

All this was all the more tempting because the doctrine still allowed for many kinds of variant. We have already spoken of the Jesuit mishmash and the Leiden 'new-old' philosophy, but the whole of Europe swarmed with such compromises. Furthermore, for any given natural phenomenon you were not obliged to follow Descartes but were free to imagine some alternative mechanism. You could water down the strict distinction that his early followers maintained between empty space with atoms (as Beeckman and Gassendi had done) and space filled with whirlpools of matter (Descartes himself). You could approach the absence of movement differently from Descartes by adding to his laws of motion a separate principle of rest. You could even start worrying about whether there were any limits at all to our freedom to imagine any kind of particulate mechanism. Or was everything permitted, did arbitrariness rule the roost and was Isaac Newton right when he complained despairingly in 1679 about 'natural Philosophy, where there is no end of fansying'?

In essence, Newton was right, yet we can also see that adherents of this type of natural philosophy did apply certain standards in practice. Which corpuscular mechanisms were or were not permissible? From the kind of explanation that they came up with in their day-to-day philosophising, it is possible to extract four criteria which they applied to their speculations.

Above all there was the certainty provided by the first principles. For Beeckman and later thinkers in the same spirit, that certainty lay in the requirement that you must be able to visualise a certain mechanism – it needs to be 'picturable'. An 'occult' force or a force working in sympathy is impossible to imagine in concrete terms, quite unlike an effect brought about by the successive rarefaction and condensation of the air. This criterion of 'picturability' was widely shared among the adherents of 'Athens-plus'. Descartes himself had based his system on a clear and distinct chain of reasoning that had led him from the *cogito* to his whirlpool world.

Closely connected with this was the demand for consistency. This also worked primarily at the level of first principles. Descartes even prided himself on the systematic nature and the clear-cut order of his deductions; the whole chain of reasoning had to form a single, orderly whole. At the level of separate phenomena, by contrast, you could interpret them now in this manner, now in that, and still be fully consistent with the first principles. For instance, Beeckman could at one moment make musical pitch dependent on the *speed* of the particles which a vibrating string supposedly carves out of the air, and at another time on the *size* of those particles. But was consistency with the first principles not on occasion more apparent than real? Critics of atomism had forever wanted to know how those particles manage to join together in temporary conglomerations. What special 'glue' holds them together for a while? Stoics had a ready answer – they knew perfectly well that *pneuma* was what provided the needed cohesion – but the atomists did not have any *pneuma* in stock. Extraordinarily subtle particles of matter were then called in to do the job, and in the generation after Descartes we can see how, in the hands of particle thinkers, these extra-fine particles and Stoic *pneuma* creep ever more closely together, leaving less and less of the desired theoretical consistency.

A third criterion was analogy. It was very different from the analogies employed by the mathematical scientists. They used them to see how far you could get by applying a mathematical regularity known for one type of movement to another one as yet unknown but with luck sufficiently similar. In natural philosophy, in contrast, analogies were being drawn all the time between the micro-world of particles in motion and the macro-world of daily phenomena. When Beeckman wants to explain how his sound particles affect the eardrum he literally compares it to a drum roll that is faster or slower and performed with larger or smaller drumsticks. Analogy from the macro-world is really the only reality check that was available to this kind of particle thinking. After all, even under the microscope, particles and their movements are too small to be visible. So the reality check offered by analogy is very limited – analogy makes it easier to imagine a certain micro-mechanism in a concrete sort of way, but that is all there is to it.

In practice, there is a fourth criterion which puts a brake on unrestrained fantasy. It is a very informal concept that I introduce here; the modern term for it is 'physical intuition'. Take Descartes' opinion of Beeckman's notion that sound is transmitted by the particles which the vibrating string cuts out of the air and sends on their way. 'Ridiculous' was Descartes' concise assessment. Now what does that mean? Beeckman's idea is certainly in accordance with the first principles of his natural philosophy, it is consistent with the rest of his conception of sound, the analogy with the visible world is obvious, and yet it is completely misconceived and Descartes rightly sensed it. There were nonetheless occasions when Beeckman, as also Descartes, was very perceptive in explaining how a certain mechanism works. Descartes saw the transmission of sound as wave-like; Beeckman perceived that dissonance is due to beats. They formulated these insights in ways that do not, indeed could not, get it entirely right. But what clearly emerges from them is that unlike so many of their successors in the next generation they did not simply philosophise their insights out of thin air. Both men certainly possessed something in the nature of 'physical intuition' – a far from infallible yet at times perceptive insight into underlying patterns and how they operate, which is taken nowadays to distinguish the great scientist from the average.

Even these four criteria together could not definitively prevent fantasy from running riot, as practitioners of 'Athens-plus' were often prone to do. In the generation following the pioneers, as well as the usual herd of scholarly windbags and armchair philosophers, there is also the occasional more disciplined thinker searching for additional criteria. One obvious potential aid was sought in mathematics. A mathematician from the school of Galileo, Giovanni Alfonso Borelli, attempted to subject particle mechanisms to the test of geometrical form. For instance, by no means every geometrical form lends itself to being stacked. So if a particular natural phenomenon, e.g. the freezing of water, seemed to be explicable by the stacking up of particles, those particles would have to take the form of, for example, a prism or a hollow cone; other candidates could now be excluded. The mechanisms which Borelli devised in his study of such phenomena cannot but appear to the modern physicist as the height of absurdity. Yet underlying them was a healthy perception that mathematics provides the best means of imposing discipline on the power of the imagination which the scientist really cannot do without.

A contemporary of Borelli's who took that healthy insight forward but with rather greater competence was Christiaan Huygens. His father Constantijn was a good friend of Descartes, and Christiaan had read the *Principia philosophiae* as soon as it came out. Half a century later, towards the end of his life, the 64-year-old Christiaan wrote down his recollection of the experience:

> The novelty in the figures of his particles and whirlpools make[s] it a great pleasure. When I read this book of Principles for the first time it seemed to me that everything in the world ran as well as it could, and when I found some difficulty in it I believed that it was my fault for not grasping well what he meant there. I was only 15 or 16 years old.

In the passages which immediately follow, Huygens effectively dismantles this initial impression. He pulls to pieces just about all the non-mathematical insights that Descartes had ever published as well as the 'all or nothing' style of thinking behind it, observing rather unkindly that 'Mr Descartes had hit upon the way to have his conjectures and fictions accepted as truths.' At an early stage,

Huygens himself had worked out ways and means for taming Descartes' corpuscular fantasies through mathematics (we shall see how in the next chapter). Nevertheless, from the tenderness with which the older Huygens looks back on his youth we clearly sense the fascination he had once felt for Descartes' total system. And indeed whatever aversion a mature person might, indeed should, feel for such an uncompromising, total mental construct of the world seen from a single self-obsessed viewpoint, anyone who has not experienced the attraction of such a vision in their youth was born old and will never achieve anything truly great. The market for obsessive, total systems never dries up, be it Rousseau's *Du contrat social*, Marx's *Kapital* or Descartes' *Principia philosophiae*.

The question remains why among the countless attempts at monomaniacal total systems only these and a few comparable works have managed to attract such wide support. Support, and this is so remarkable, that can fall away as quickly as it arises. Are we dealing here with just the snowball effect that was so clearly at work in the seventeenth-century particle mania? The way in which only a few such visions, as distinct from all the others, succeed in exercising such a massive effect is for us historians one of the great puzzles that the past poses. In his own, slightly ironic manner Christiaan Huygens has given us a hint. When rounding off the critical section of his comments he writes that Descartes

> has with great ingenuity fabricated ... this entire new system, and given it such a turn of truth-likeness as to make infinitely many people satisfied with it and pleased with it.

'A turn of truth-likeness' indeed! From time immemorial the human mind has opened itself to the temptation of systems which provide the right answer for everything. What Huygens is suggesting here is that Descartes was one of those rare writers able to present to the reader a personal take on his world as ineluctably valid – at least for so long as it lasts.

EXPLORATORY EXPERIMENT GAINS GROUND

The ideological neutrality of experiment did much to encourage its use after about 1660 among the Jesuits, as well as in the Académie and above all in and around the Royal Society. Another cause of the rapid expansion of exploratory, fact-finding experimental research was the Baconian expectation of practical applications whether or not they actually came about (usually not). And finally there was a powerful stimulus in the search for facts itself. The realisation that the world is much richer in natural phenomena than our senses can perceive had now taken hold and so, in principle, an unlimited field of research had been opened up. The big problem was not so much what topic to choose from the plethora available, but how to impose some order on all the hidden phenomena and their properties that nature now appeared to have on display. In quite a different way from the philosophy of moving particles, the question was here, too, how to curb the arbitrariness that was lying in wait, an arbitrariness resulting from the random accumulation of data without direction or structure.

Not that it always worked. In particular, the system of lists prescribed by Bacon appeared not to work at all. He had called a list of phenomena around a core subject a 'natural history'. As the century advanced many a scholar embarked on such a 'natural history' of colour, of air or even of the arts and crafts in their full breadth. Rather fewer of them were able to complete such bulky studies. There were more fields where the sheer feat of collecting threatened to be disrupted by the immense quantities of material. Collections were built up and put on display in private museums, and to impose some control over the mass of curiosities catalogues were written up, many of which had still not been completed when the collector passed away. The unceasing arrival from the tropics of unknown shrubs, plants and herbs marked the end of the current classification of the plant world – the search for a new system was to culminate in the eighteenth century with Linnaeus.

With specific-experimental research into nature's hidden properties it proved a little easier to cope with the constant threat of disorder. The research seemed to group itself of its own accord around four new instruments and the four or five subjects which the pioneers, Gilbert, Harvey and van Helmont, had first opened up by means of systematic experimentation. As well as light (see the next chapter), these subjects were magnetism, electricity, physical and mental health, chemistry and alchemy. The new instruments were the telescope, the microscope, the air pump and musical instruments. The latter were of course not in themselves new. What was new was that violins or organs or trumpets began to be used to investigate the natural world. And the air pump was such an intriguing and spectacular piece of apparatus that it would become a symbol for the innovative ways of investigating nature that held Europe in thrall in the second half of the seventeenth century.

At first, the air pump was the means by which Otto von Guericke, mayor of Magdeburg in Germany, made use of Torricellian space. A mercury tube does not lend itself well to experimental work – it can be used for little else than the barometer which it soon became after Pascal. This is so because the equilibrium between the mercury column and the column of air that lay at the basis of Pascal's experiment on the Puy-de-Dôme can also be used the other way round. The rise or fall of the mercury corresponds to higher or lower air pressure so, if you make the mercury level measurable by putting a scale on the tube, you can determine changes in the air pressure. That is why, at home, you can now nail a well-reasoned weather forecast to the wall. But you can do a great deal more with air pressure. You can make measurable the vast forces that are hidden within it. You can demonstrate them theatrically. You can seek to tap them for general use. Von Guericke aimed for the first two objectives. His famous demonstration with the 'Magdeburg hemispheres' was repeated a few decades ago in Zaandam in the Netherlands. Once again, two teams of eight horses each did not succeed in pulling apart a pair of well-fitting brass half-globes once the air had been pumped out of them.

Von Guericke's demonstration with the Magdeburg hemispheres

On reading von Guericke's publication, it occurred to Christiaan Huygens how it might actually be possible to tap the giant forces apparently latent in the atmosphere. The case was typical of Huygens. Ideas used to come to him on learning of those of others, followed as a rule by a tacit 'I can do that better.' In this case, the idea he came up with was the design of a gunpowder engine.

In this form the design would have proved unworkable if Huygens had ever built it. His laboratory assistant in Paris, Denis Papin, kept the metal cylinder, placed it over a flame and replaced the gunpowder with water. He surmised quite correctly that, if the cylinder was suddenly cooled down, the steam would condense and in so doing produce a vacuum. But this improved version proved unworkable, too, until early in the eighteenth century a blacksmith, Thomas Newcomen, overcame the practical difficulties with some brilliant technical solutions. Newcomen's 'fire engine' helped to keep

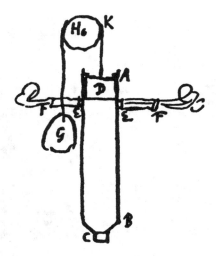

Huygens' design for a gunpowder engine

Container C is filled with gunpowder and a burning wick and attached to cylinder AB so as to close it off. The explosion will, according to Huygens, drive the air out of the cylinder into the wet leather tubes EF. The cylinder is now empty, the tubes are also empty and hang limply against the cylinder. Air pressure takes over and presses the piston D to the bottom of the cylinder. This causes weight G to be raised via pulley K. That is the working stroke. According to Huygens, the working stroke could be used for making fountains squirt, or for erecting obelisks or for turning wheat mills in places unsuitable for horses, with the further advantage that an idle gunpowder engine does not eat oats.

Britain's mines dry for over half a century until in 1765 James Watt radically altered the design due to an even more brilliant blend of, on the one hand, engineering ingenuity and, on the other, scientific insight, this time chiefly of his own making. Thus came about the steam engine, which soon became the universally adaptable force which gave a decisive impulse to the Industrial Revolution. And so the Baconian dream was finally realised after surviving for over a century mainly as an ideology. But after this brief foray into the future, we must return to the fortunes of the air pump in the seventeenth century.

Besides the demonstration and the hoped-for practical use of atmospheric pressure, the void was also, indeed mainly, used in experimental research. As well as von Guericke and Huygens, those who threw themselves into it most intensely were Robert Boyle and his assistant Robert Hooke, who later as Curator of Experiments helped steer the Royal Society away from descending into all too trivial pursuits. They built an air pump which eventually proved good enough to pump air out of a glass sphere down to around 2 cm mercury pressure (as estimated in retrospect).

Though not a complete vacuum, then, it was quite a good approximation. All kinds of objects were placed in the evacuated receiver to see what would happen. Some behaviour was highly predictable. Few people were surprised that animals placed in the container soon gave up the ghost which according to Descartes they did not even possess, or that the sound of a ticking clock died away as the pump did its work. More exciting was what might happen to Robert Hooke when on 23 March 1671 he spent some fifteen minutes in a space that had had a quarter of its air pumped out. He emerged as fresh as a daisy, only complaining about some earache. Completely unpredictable, meanwhile, was the fate of two substances which were heated in the void space from the outside. Reintroduced to the air, a ruby promptly disintegrated into ash, while a lump of butter was literally left cold by its adventure.

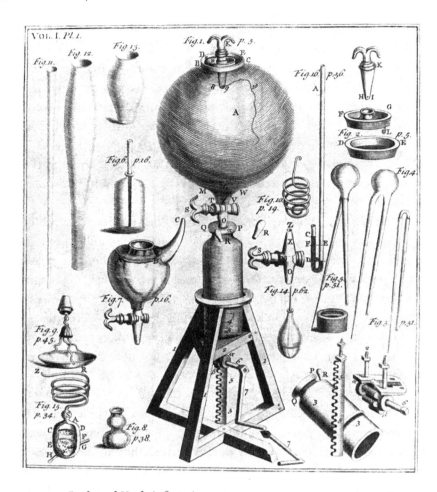

Boyle and Hooke's first air pump

Extensive 'natural histories' were filled with this kind of outcome. Extensive indeed, as the Royal Society imposed norms of its
own for acceptable ways of reporting. In particular, the secretary
Henry Oldenburg, who also owned and edited the journal *Philosoph-*
ical Transactions, kept an eagle eye on such things. A report had to be
businesslike, without any rhetorical tricks, flowery turns of phrase or
deliberate human touches. 'Not words but things' was the slogan.
And the report must be so detailed that anyone who did not attend

the experiment in person (in other words, almost everybody) would be convinced that it had all happened in the way described and not in any other. That is very different from the extreme brevity of the reporting of validating experiments in mathematical nature-knowledge. In contrast to the detailed report that Pascal received from his brother-in-law, he habitually reported on his other vacuum experiments in roughly the following way: 'If you do this, it leads to that, just as my theory already led me to expect.' This kind of description was much more directed towards general cases than any specific one. Such a style of reporting awakened in Boyle a doubt (in a few cases quite rightly so) about whether Pascal had actually carried out his experiments on the vacuum as described. That is why Boyle was wont to address those absent in phrases something like the following: 'Under the locally prevailing circumstances, which I shall now list in detail, I saw this happen, then that, and Lord X and Mr Y Esq. saw it, too. And oh yes, may I suggest the following by way of a possible explanation of how all this may perhaps have happened in the way that it did?'

With that we arrive at the question of what if any coherence could be detected between the hundreds of experiments, each one reported in so confidence-inspiring a fashion. How did the experimenters of the Royal Society impose order on such a multiplicity; how did they decide which assertions were valid and which were not? In essence it is the same question that we considered earlier in this chapter in connection with the two other revolutionary modes of nature-knowledge. It is even the key question of all later research in the natural sciences. What makes this period special is that the question presented itself for the first time in so compelling a manner. The modern natural sciences possess an arsenal of means – not infallible but certainly effective – for distinguishing claims that hold water from those that are plausible at best. The search for means of that kind began in the seventeenth century *and was different for each of the three new modes of nature-knowledge*: mathematical, natural philosophical and fact-finding experimental. In the latter case, I shall no longer confine myself to the void, but pose the question across the whole range of fact-finding experimental research. And to find an answer we shall need to look at the numerous obstacles that had to be overcome as well as the few resources which could be called upon to help.

The greatest obstacle lay in the capriciousness that nature so often displays in experimental research. By that I do not mean exactly the 'untidiness' due to which in mathematical nature-knowledge a simple model often proved insufficient. 'Capriciousness' goes a step further to include phenomena which initially at least the researcher is completely at a loss to make sense of, as for instance the contrasting behaviour in a vacuum of the heated ruby and the heated lump of butter. Indeed, things could at times get even worse:

> The malevolence of inanimate objects is nowhere better instanced than in the phenomena of frictional electricity. Their apparent caprice consistently frustrated the efforts of early theoreticians trying to reduce them to rule. Consider the effect of moisture on the surfaces of insulators and in the air surrounding them. The early electricians realized that contact with water enervated an otherwise vigorous electric, like amber, but they did not fully recognize the effect of humidity. On a sultry summer day, or in the presence of a sizable perspiring audience, experiments that had often succeeded might suddenly and inexplicably fail; while the operator himself, sweating at his task, helped to dissipate the charges he intended to collect.

Natural phenomena might not only turn out to be capricious, but they could also be plagued by impurities. Looking back on many a chemical reaction we can be fairly certain in retrospect that the substances used must have contained impurities that affected the results. Also the instruments with which much experimental research was carried out suffered from a whole range of shortcomings. When we consider the microscopic investigations of the great men of the time, such as Hooke, Leeuwenhoek and Swammerdam, one can only admire the inventiveness and skill with which they managed to illuminate their objects for study with no other means available than sunbeams or a candle. They were particularly skilful in learning to distinguish between genuine phenomena and such misleading effects as were brought about, for example, by a tiny scratch on the lens. And the greatness of a telescopic observer like Cassini lay not least in his systematic efforts to eliminate sources of observational error. He even sent an assistant to the ruins of Tycho's Uraniborg to measure its precise latitude and longitude using the more accurate instruments that he had built in the meantime. Only thus could Cassini compare with the required accuracy his own observations of stars and planets with Tycho's from a century earlier.

Where there were so many obstacles it is not strange that experimentalists were generally reluctant to provide explanations of what they observed. Their reluctance was strengthened by two other considerations. One we are already familiar with: to provide explanations had always been the preserve of the natural philosophers, and they were now out of favour with the Académie and the Royal Society. Experiment was embraced precisely because it did not suffer from the ideological baggage that had always been a characteristic of natural philosophy. One more reason for reluctance rested in an acute awareness of the bias to which every researcher is inevitably subject. Francis Bacon had devoted some compelling passages to the topic. The human mind, he pointed out, is not a smoothly polished mirror that reflects reality without distortion. It is 'rather like some enchanted mirror, full of superstitions and ghosts'. Therefore we must be on our guard all the time against our inclination to jump to conclusions and bridge in one bold leap the gulf between observation and generalisation, or to maintain forever our love affair with our own findings.

In spite of all these good reasons to draw conclusions from experimentally based observations with a certain reluctance only, most could not resist the urge to impose at least some order on the chaos. Without at least something of a theoretical guideline you rarely get very far; a theory of some kind can lead you from one experiment to the next, enabling you to build up a progressive series. With the best we encounter a kind of triad. Theories well up from an underlying world-view, followed by an interchange between experiments pursued in an orderly manner and increasing refinement or even readjustment of the original theory. As a rule, though not always, the underlying world-view is one of particles in motion, not as a fully fledged, dogmatic natural philosophy based on first principles, but in a more pragmatic format. This kind of world-view was seldom fully consistent or wholly unambiguous – in the background to this type of corpuscular thinking all kinds of magical and/or vitalist conceptions could and often did play a role, which then found their way into the theory.

The results ranged accordingly from the virtually absurd to the unquestionably brilliant. Criteria for telling them apart scarcely existed. But even without such criteria there emerged within both the Académie and the Royal Society a healthy distrust of the sometimes quite gullible accounts to which some leading Jesuits treated their audience. At the practical level the pre-eminent criterion was, of course, the ability to repeat an experiment. This was another reason why the style of reporting had to be so detailed. But even this did not always provide a decisive criterion, as we saw earlier in the case of frictional electricity. Nor did means lie ready to hand for reining in the somewhat directionless proliferation of theories. For instance, most Jesuits thought that the electrical effect produced by the rubbing of amber stems from its pores being opened thereby. Subtle matter now gets a chance to flow out, the surrounding air gets rarer, and as the air regains its original density light objects such as snippets of paper are driven towards the amber. The London-based researchers, on the other hand, stayed with Gilbert's explanation of electrical attraction due to a water-like flow of thin viscous threads. There were no conclusive reasons for choosing one explanation rather than the other, so in practice the choice would rather depend on extraneous factors such as nationality or the underlying world-view from which the explanation emerged.

One of the better results of experimental research was the discovery by van Leeuwenhoek of the millions of creatures crawling around in his sperm which he placed under his microscope after leaving Mrs van Leeuwenhoek still panting in bed (a pithy detail not left unmentioned in his report to reassure his readers with their instant recall of Scripture that he had not committed the sin of Onan). He made the discovery five years after what, in essence correctly, his Delft colleague Reinier de Graaf took to be the discovery of the egg in women. Inevitably, there followed an as yet insoluble debate between 'ovists' and 'animalculists', that is, between attributing reproductive primacy to the egg or to the 'little animals' in the sperm, respectively.

As a rule, experimental research remained qualitative. On the few occasions that quantities played a part these were usually obtained by measurement. This was not only true of the positions of the stars and planets as seen through the telescope or of the large-scale surveying project that Cassini undertook for Louis XIV. It was also the case with the size of the universe, which in less than three-quarters of a century had to be greatly enlarged. From a diameter of 20,000 times the radius of the Earth, the universe grew to an infinity in which the distances between stars could only be expressed in the days or even weeks that light was taken to need for travelling from one star to another (a unit of measurement to appear first in a 1694 issue of the *Philosophical Transactions*).

Outside astronomy, too, quantities played a role, as in the inverse relationship on which Boyle stumbled between the pressure and volume of air (still known as Boyle's law). Quantities also had their place in chemistry and alchemy, which were closely linked and jointly often called 'chymistry'. In particular, George Starkey and Isaac Newton, by carefully weighing the substances that they used, raised their experimental researches in this domain to a level even higher than that attained by van Helmont. But there were also those who were deeply suspicious of the central assumption of the alchemist that metals attain maturity within the earth and that we can hasten the process artificially. Not that the grounds for that mistrust were the same as those on which alchemy would later be banished by scientists as sheer superstition. It is just unhistorical to make a strict retrospective distinction between enlightened scientists and the superstitious masses in the seventeenth century. There were many capable researchers who believed in witchcraft or in exorcism, as there were many opponents of the new nature-knowledge who did not share these popular beliefs, most often on biblical grounds.

Three scholars achieved a particularly high level of research on the plane where theory and experiment came into fruitful interaction. The Jesuit Lana Terzi inferred from the explanation of electrical attraction which he helped to draw up the possibility of electrical *repulsion*, which he went on to demonstrate in a fine series of experiments and then to map in quite some detail. A Fellow of the Royal Society, Francis Robartes, during one of its weekly sessions, made a connection between the previously discovered multiple vibration of a string and the natural tones of a trumpet. And in far-away Bologna, Marcello Malpighi, a corresponding member of the Royal Society, achieved lasting results by carrying out subtle comparative microscopic research into how glands in humans and animals excrete various substances.

This same Malpighi also provides us with an instructive insight into the extent to which all this experimental research produced practically useful results. Towards the end of his long life, a fierce attack was launched against him by a certain Girolamo Sbaraglia. According to this young upstart, Professor Malpighi had no business in the Medical Faculty where healing the sick was the aim; instead, he belonged among the philosophers. All those experiments and theories and explanations which Malpighi had produced over more than thirty years of assiduous research had contributed absolutely nothing to the recovery of a single patient. To defend himself against this attack which threatened his reputation, his status and his income, the old man pulled out all the stops to show how useful his research, and that of his co-microscopists Hooke, Leeuwenhoek and Swammerdam, actually was. The long list which Malpighi assembled is as sad as it is instructive. If you go through the list critically and with a sober mind, without regard for your natural dislike of a cheeky non-achiever taking pot-shots at a great man, it quickly becomes apparent that Sbaraglia was right. The list shows convincingly what Sbaraglia had not disputed, namely that Malpighi had made many phenomena surrounding sickness and health more comprehensible. But only with a degree of goodwill can one pick out two fairly inno-cent kidney complaints where that understanding had led to better treatment. As such, the case of Malpighi versus Sbaraglia exemplifies the practical uses that the new modes of nature-knowledge could be put to. Even by the end of the seventeenth century there were almost none.

The following conclusions can be inferred from the above. In the course of the seventeenth century, fact-finding experimental research produced some lasting results, even though practical use was hardly one of them. The central obstacle, how to come to grips with nature in all its capriciousness, was tackled with great inventiveness, but on the theoretical level, in spite of all the precautionary measures taken, it was very rarely resolved. And even when it was, there were hardly any criteria which could provide some degree of certainty about the valid-ity of the solution obtained. On balance, here too, though in a way quite different from natural philosophy, arbitrariness appeared to reign supreme. And the question which occurred to a few investiga-tors at the time was whether anything fundamental could be done about it. The answer to that question was 'yes'.

6 Revolutionary transformation continued

The previous chapter may have given you, dear reader, the impression of being shown round three different rooms in the house of nature-knowledge. First of all you visited the kitchen, then the best room and then the playroom. In each room the furniture was very different. Each mode of nature-knowledge differed not only in its outcomes but also in its style of reasoning, in its handiwork (if any were involved) and in its exploration of ways and means to counter what Newton called 'fansying'. In other words, each of the three worked largely in its own way, and the high walls separating 'Athens' from 'Alexandria', which we first encountered with the ancient Greeks, were still standing unscathed. Although in Europe, together with a third mode of nature-knowledge, they had gone through a revolutionary transformation, hardly any effort had as yet been made to end their mutual isolation. It is true that we have encountered a few individuals, especially Huygens and the young Newton, who were involved in both realist-mathematical nature-knowledge and exploratory, fact-finding experiment. But for each mode they wore, as it were, a different hat. They behaved in the one quite differently from in the other and in each they followed prevailing practices and styles of thought (which they seized upon to make some splendid discoveries).

And yet it was also these two, Huygens and the young Newton, together with Boyle and Hooke, who managed to some extent to break down the walls separating different modes of nature-knowledge or at least to make sizable breaches in them.

BARRIERS BROKEN DOWN

Three aspects to this breaching of the walls should be noted. There is the political and religious background which we might call 'the spirit of Westphalia'. Then, within each of the three modes of revolutionary nature-knowledge there are certain recurring themes which more or less incidentally lend themselves to crossover. One such theme is the making of measurements, the other is thinking in terms of moving particles. Finally, there is the unprecedented fact itself of crossing the boundaries between distinct modes of nature-knowledge. It takes the form of two new, and again revolutionary, transformations. For certain specific subjects such as impact, circular motion, light and chemical reactions it proves possible to merge the practices and styles of reasoning current in two distinct modes of nature-knowledge. Such partial mergers were made possible by one grand, decisive action – to rid the philosophy of moving particles of its dogmatic knowledge structure and conceive of it instead as an open-ended, potentially fruitful hypothesis. This transformation from dogma to hypothesis is the turning point and Huygens is its pioneer.

All this sounds rather abstract so let us now look at how it all came about in practice.

The Peace of Westphalia in 1648 and the English Restoration of 1660 opened up the prospect of a radical turn-around. People were no longer quite so keen to push disagreements to extremes; on the contrary they sought to tone them down. Reconciliation and a greater readiness to accept some compromise came to dominate European attitudes. The new spirit involved politics and religion in the first place, but soon spread to incorporate nature-knowledge with which they were so closely connected. In the past, only a few individuals had displayed such a conciliatory spirit. A notable example was Marin Mersenne, a Friar Minor. Between about 1625 and 1648 he exchanged letters from his monastic cell in Paris with almost everybody in Europe who had an interest in innovative nature-knowledge. He visited Beeckman in Dordrecht and was allowed to browse the latter's scholarly diary. He was friendly with Descartes, who used him as an inexhaustible source of information about the latest Paris gossip, in particular gossip about himself. He tried, though in vain, to extend his network of correspondents to include Galileo, whom he greatly admired. Peace-loving and genuinely interested in others, he must have been a sweet-natured man, a breath of fresh air among so many self-obsessed, hot-headed megalomaniacs. Mersenne's own, at times skilful, research into world harmony combined rather inconsistently the mixed mathematics that he had acquired at school from the Jesuits with experiments that were largely fact-finding though occasionally for testing, all enriched with calculations wherever he felt any need for them. However, he had no time for those ceaselessly moving particles that populated the universe in the view of his host Beeckman and his friend Descartes. So he quietly stripped away the underlying atomist mechanisms from the swathes of musical theory that he borrowed from Beeckman.

Such pragmatic handling of widely different approaches became rather the rule in the years following the Peace of Westphalia and the Restoration. Disputes between natural philosophers and those innovators who trespassed uninvited on their patch now lost their bitterness. Hobbes, a thinker still completely in the dogmatic mould of classic natural philosophy, fought a rearguard action against Boyle with his own variant of particle thinking on the question whether there really was a void under the glass bell of the air pump. The point was that Hobbes' first principles excluded the possibility of void space. Like Galileo's opponents in Pisa half a century earlier, Hobbes was unerring in his ability to pick out the weakest part of his opponent's position. He argued, not entirely without reason, that Boyle's air pump leaked like a sieve, and he exploited with glee a typically capricious outcome of one specific experiment to drive his point home. But in Restoration England there was little patience left for that kind of dogmatic quibbling, and Hobbes remained an outsider, always under suspicion of actually being an atheist – in half a century the tables had definitely turned!

One aspect of the new nature-knowledge where compromise and reconciliation in the guise of a certain amount of crossover were realised was measurement. This may sound surprising to those who think that mathematics and quantity boil down to more or less the same thing. Certainly Galileo regarded the world as the embodiment of mathematics, but that did not mean that he had any great interest in actual quantities. He derived the relationship between speed, time and distance in vertical descent but he never stopped to wonder how far an object actually falls in a second. Among those who, on reading Galileo, did want to find this out to the last minute detail were Father Riccioli SJ and Brother Mersenne. Both of them laboured away on hugely time-consuming experiments with pendulums, but the precision of their results was more semblance than reality. Even so it was from their detailed records and complex calculations that Christiaan Huygens became acquainted with the problem of the distance covered while falling for 1 second. With his training in Galilean thinking he at once dragged the problem out of the domain of mixed mathematics and fact-finding experiment back to realist-mathematical nature-knowledge. It was this modest problem that marked the start of his three-month-long whirlwind of discovery. What since Newton's *Principia* is called the gravitational constant came as a neat byproduct of one of Huygens' greatest discoveries. So, halfway through the seventeenth century, mathematical proportion and the determination of quantities began to come together. Perhaps the clearest illustration of this is in telescopic astronomy. Quite soon after its emergence from the crafts, the telescope became chiefly an instrument in the service of mathematical nature-knowledge. It occurred to Huygens to equip it with a finely graded scale, which is how in the 1660s the telescope turned into an instrument of measurement.

Particles in motion lent themselves even better to crossover of this kind than measurement. We encountered this briefly with Borelli, and also as a background world-view when fact-finding experimenters wanted to explain this or that experimental result. Leeuwenhoek, worthy husband and gifted observer but far from subtle theoretician, invoked particles all the time to explain what he observed through his wonder lenses. He stuck, as it were, particles all over his observations, but to little avail. Borelli, in contrast, had started out with particles in motion and then became worried about the apparently boundless freedom they left to fanciful mechanisms. Geometric form, as we saw, seemed to him to offer a means to impose some discipline in this regard.

And with that we arrive at a very important point, namely the high level of arbitrariness to which both moving particles and exploratory experimental research were exposed. Attempts to rein in that arbitrariness did not get very far, as we saw in the previous chapter. Consequently, the revolutionary idea arose of stripping particles in motion of their more fanciful features by binding them, not loosely but firmly, to either realist-mathematical or exploratory-experimental nature-knowledge. That it was at all possible, and how to achieve it, was perceived only by the greatest minds of that generation: Huygens, Boyle, Hooke and the young Newton. Huygens and the young Newton hit upon the first possibility, to rein in particles in motion through mathematics. Boyle and Hooke and the same young Newton saw the possibility of reining them in through exploratory experiment. And so, after the three revolutionary transformations between about 1600 and 1645, two more took place between about 1655/60 and 1684.

In this quartet of second-generation revolutionaries I have now referred more than once to 'the young Newton'. This is to distinguish between the two great creative periods in his life. From 1665 to 1668 (his famous wonder years), unwittingly in parallel with Huygens, he brought about the fourth revolutionary transformation. From 1669 to 1679 he took part in the fifth, exchanging views at times with Boyle and Hooke. And between 1684 and 1687 the now fully mature Newton brought about a sixth and final revolutionary transformation, entirely on his own, with his laws of motion and of universal gravitation as the most impressive outcome. We shall now consider these three transformations in turn.

MATHEMATICAL SCIENCE ENRICHED
WITH CORPUSCLES: HUYGENS AND THE
YOUNG NEWTON

An indispensable precondition for coupling the philosophy of particles in motion firmly to one of the other two modes of nature-knowledge was first to strip it of its 'Athenian' knowledge structure. The time-honoured claim of the natural philosopher to omniscience and indisputable certainty had first to be peeled away. The man who accomplished this feat between 1652 and 1656 was Christiaan Huygens and the immediate trigger was the problem of collision.

Problem? What problem? Had Descartes not already both posed and solved the question of how objects collide in *Principia philosophiae*? Certainly he had, and at the age of 15 or 16 Christiaan had swallowed this unquestioningly along with all the rest. But a year or so later, as a student at Leiden, he had been introduced to mathematical nature-knowledge. He had first begun to pursue this in 'Alexandrian' fashion, with virtuoso refinements of the work of Archimedes, but before long Galileo's study of motion drew him permanently into the 'Alexandria-plus' camp. When in 1652 at the age of 23 he again immersed himself in Descartes' rules of impact from this new perspective, a closer inspection yielded some remarkable findings. For example, Descartes argued that, if a smaller sphere collides with a larger one, it will rebound with its velocity unchanged, whereas the larger sphere will just stay put. Try that out with two balls suspended from strings, and you will see that the larger one definitely begins to swing. I once helped mount an exhibition on Huygens at the Museum Boerhaave in Leiden. We wanted to enable our visitors to distinguish between Huygens' and Descartes' rules of impact for themselves. The only way to demonstrate the latter was by attaching the large billiard ball to the wall with a rod.

Whatever else Descartes may have been, he was not stupid. He had derived his rules of impact in true Athenian fashion from his own first principles – hence their *a priori* certainty. But at the same time he was well aware that most of those rules fail when applied to billiards. Only, in his view this did not matter at all. In his whirlpool world objects collide without cease yet never in a pure form, since all particles are in motion all the time. In the Galilean approach you can, indeed you must, think such disruptions away. But in Descartes' world the vortices cannot be 'thought away'. Far from being disruptive phenomena from which we can abstract, they are the very core of the natural world. Billiards teaches us nothing.

Huygens spent four years, on and off, working on the problem of impact, and during that time he wrestled with a dual loyalty. There was the overwhelming impression that Descartes' *Principia philosophiae* had made on him earlier. And there was his recently formed conviction that Galileo had hit upon the way to make mathematical nature-knowledge deal with real natural phenomena. Galileo had not explored the question of what happens when spheres of varying sizes coming from different directions collide with each other – that question is much more obvious in a world of particles in motion. Nevertheless, Galileo's general approach to motion had clear-cut consequences. Considered in the framework of 'Alexandria-plus', billiards definitely provides an experimental test – one which Descartes' rules simply fail to survive. Of course the rolling of ivory balls, however hard and smooth they are, across baize, however smooth it may be, will never correspond exactly to the theoretical outcome. To that extent Descartes was right, so Huygens acknowledged. But he went on to argue, following a very innovative line of reasoning, that if you have two sets of rules, and then find at the billiard table that one set runs completely counter to experience while the other produces a fair approximation, that is surely a powerful indication that the first set is definitely wide of the mark while it is fair to assume that with the second set you are at least on the right track. And so we see Huygens on his way towards an important refinement of 'Alexandria-plus'. Certainly the outcome of an experimental test cannot fail to deviate a little from what the theory predicts, but you can in turn try to incorporate the deviations themselves into a more refined mathematical model.

It was not only this question of whether deviation was admissible that drove Huygens away from Descartes. A core element of Galileo's account of motion is its relativity. We discussed it earlier with the example of the two trains: is the one at rest, or the other? The relativity of motion implies that it makes no difference whether a small sphere collides with a large one or a large one with a small one. The two cases are identical; it is only the frame of reference that has shifted. That is precisely what the relativity of motion is about. But in Descartes' rules of impact, shifting the frame of reference does make a difference. On this point Descartes' shortcomings emerged even more clearly, since he himself had actually posited a principle of relative motion, but then had momentarily lost sight of it in deriving his rules for colliding bodies.

This led Huygens, with the help of the self-same principle but now applied consistently, to go in search of new rules of collision that would also satisfy the billiards player. Apart from their mathematical form of expression, they are still taught in school physics today. He took as his starting point the specific case of two differently sized billiard balls that approach each other at such a speed that the product of weight and velocity is the same for both. In that case each of them will rebound at the same speed that it had before impact. If you now provide the space in which the balls collide with an arbitrary velocity, you can successively work out every other conceivable situation. Because the balls move away from each other in this one specific case, they will do so in all cases, and at exactly the same speed. In the Museum Boerhaave we had no need to screw the large ball to the wall; in collision with the small one it came sufficiently close to doing what it was supposed to do.

The way in which Huygens deduced his own rules of collision involved more than consistently applying the principle of relativity of motion. At first he had tried to derive his rules by means of a 'collision force' – the idea that at the exact moment when the motion is being transferred, one ball exerts a force on the other that drives it back. But he soon gave up on that idea. On reflection he found 'force' too obscure a notion to risk burning his fingers on it. It was too loaded with the occult forces of attraction and repulsion that figured so prominently in the magical world-view that Descartes had gone out of his way to eradicate with his whirlpool world. In this respect, but in this respect alone, Huygens always remained a follower of Descartes. For the rest he broke with Cartesian doctrine, and it happened during his research into colliding objects. Huygens now no longer claimed to know precisely how the world is constructed. He no longer subscribed to first principles of the 'Athenian' type. In broad outline he knew that the world is made up of particles that behave in accordance with certain laws of motion. Those laws must be formulated mathematic-ally. But neither those laws themselves nor the exact mechanisms by which they manifest themselves can be derived from a preconceived idea about how the natural world is organised, and definitely not with any certainty. We can claim no more than 'probability' for our con-clusions about this or that mechanism.

These were radical, even revolutionary, tenets, but Huygens addressed the matter in such a 'business-as-usual' manner that the huge historical rift has never been noticed before. Things had not previously taken such a course. Whenever a natural philosopher had also been a mathematician, like Avicenna (Ibn Sina) or Descartes himself, he kept the two functions quite strictly separate. In the case of Huygens, during his study of objects in impact, a fusion takes place, in that he now weaves together components of 'Athens-plus' and 'Alexandria-plus'. If you consider in a comparative manner the history of natural philosophy from Plato onwards, you find that no one had ever made so free with it. Natural philosophies had at times been put through the blender, all the while retaining their speculative-dogmatic knowledge structure. Certain portions of natural philosophy had been used on occasion to flesh out an underlying world-view (as with Harvey or Gilbert), or to fill gaps in an otherwise mathemat-ical train of thought (as with Ptolemy or Copernicus). But now for the first time a whole natural philosophy is being used as a hypothesis, the viability of which is not assumed *a priori* but must be subjected to testing again and again. Nowadays we hardly know otherwise; before 1652–6 not even the possibility had been contemplated.

All this opened up something in the nature of a programme of work for Huygens. Descartes' approach was incompatible with Galileo's not only in respect of collision but for every aspect of motion. Huygens began to see it as his task to reconcile the two, not in the superficial manner in which the Jesuits had haphazardly thrown together their intellectual mishmash but by starting with particles à la Descartes and treating them mathematically à la Galileo.

With this piecemeal usage of particles in motion, from 'à la Descartes' to 'à la carte', Huygens achieved some great successes. They are among the greatest achievements of seventeenth-century research and include part of his work on the pendulum and also his success in establishing the proportions that jointly determine motion in a circular orbit.

He carried out a large part of this work in the 1650s, entirely on his own. Apart from an occasional letter which met with no response, and the clocks which formed the tangible product of his efforts, Huygens' work of revolutionary transformation remained in his study in The Hague. He was not involved in any similar activity until he was invited in 1665 to head the research activities of Louis XIV's Académie from which dogmatic Cartesians such as Rohault had so effectively been excluded. In the same period, Huygens was also elected a Fellow of the Royal Society. In a sense he had anticipated the Westphalian spirit of reconciliation and of liberation from dogmatic philosophy even before the two academies took it up in their handling of nature-knowledge. From now on, that spirit could spread its wings.

In his mission of reconciliation Huygens encountered unsuperable difficulties on only one point, and that was falling bodies. What specific particle mechanism could account for uniform acceleration, that showpiece of Galileo's *Discorsi*? Since Newton (the 'mature' Newton) we know that that reconciliation cannot succeed without introducing a very specific, very exactly defined concept of 'force'. But Huygens had rejected for good any notion of forces operating outside equilibrium situations. So the paper that he gave on the subject to his colleagues in the Académie was doomed to failure. Not only was it a failure in retrospect; one of his colleagues immediately launched a fierce critique. Gilles de Roberval insisted that without some notion of forces of attraction you would get nowhere. For once, the ever courteous and obliging Huygens found no better way of warding off this attack than bluntly to repeat his own arguments.

Roberval himself had only a vague idea of these forces. They seemed to be uncomfortably close to the occult forces that card-carrying particles-in-motion thinkers such as Huygens had steered well clear of. But meanwhile, a brand-new graduate from Cambridge named Isaac Newton was also engaged in a kind of combination of the two approaches. In total isolation he threw himself into precisely the same questions that Huygens had pored over a decade earlier. Except that Newton did so using a concept of force of his own making that he tried to subject to mathematical measure and rule. On the issues of collision, the pendulum and circular orbits, Huygens and Newton came to the same conclusions and did so without either of them knowing anything of the other's work. Even in Newton's own university, where Aristotle still reigned supreme in the 1660s, literally no one had an inkling that within eighteen months he would work his way up from being a largely self-taught beginner to the very summit of Europe's innovative natural research. Huygens' reputation was already well established but precisely in this area he had not yet published anything. And so it was bound to escape the young graduate's notice that he was drawing equal with the recognised leader in the field and on one particular point would even move ahead.

Newton's explorations did not last long. In 1668 he gave them up, as he had run into an impasse. His effort to reconcile the lines of Descartes and of Galileo using a mathematical approach to 'force' suffered shipwreck, as he himself was all too aware. Along the way, however, he had taken a step forward that had not occurred to Huygens. When he had worked out his theorem for circular orbits, he used it to test an idea that occurred to him in his mother's orchard. He was sitting there quietly when an apple fell to the ground and suddenly a light went on. Could not whatever it is that makes an object fall to the ground on Earth also hold the Moon in its orbit around the Earth? And the planets in their orbit around the Sun? Newton turned to Kepler's third law to check his calculation of an action that diminishes in proportion to the square of the distance.

Anyone who thinks that we have now touched on Newton's law of universal gravitation would be right, but also wrong. 'Right', in that from a purely mathematical viewpoint Newton had now stated one of the two parts of that law. 'Wrong', in that the quantitative check did not turn out to be entirely satisfactory. The proportion 'answered pretty nearly', and therefore not with sufficient precision for the perfectionist Newton. So like all his earlier work, these notes, too, disappeared into the drawer. But also 'wrong' in that Newton at this stage was not thinking in terms of an attractive force at all. For that he was still too much of a Cartesian, or at least a particle thinker. Until he could create mental space in his corpuscular thinking for a generous, solid concept of force that could also encompass both attraction and repulsion, the leap from his force of impact to a force of attraction would remain out of reach. In the end it was his alchemical investigations that expanded his horizons and inspired him to take that leap. And the mindset in which he did this came from a fifth revolutionary transformation which was brought about by mixing experiment with a dash of moving particles and brewing something new.

THE BACONIAN BREW: BOYLE, HOOKE AND
THE YOUNG NEWTON

The natural philosophy of particles in motion was a product of main-land Europe, of Beeckman, Gassendi and above all Descartes. The first Englishmen to become acquainted with it were a group of exiles. Leading supporters of Charles I, having lost the battle of Marston Moor in 1644, fled to Paris – not only the crown prince, who would return in 1660 as Charles II, but also Lord William Cavendish and his entourage. Two members of that entourage were his wife Margaret and the house tutor, Thomas Hobbes. The 1640s were the very period in which the philosophy of moving particles, which Beeckman had begun to explore around 1610, finally began to get through to the public. The Cavendish family drank it in, particle by particle. Hobbes and Duchess Margaret developed their own variants and set about publishing them after their return to England half a dozen years later.

Because the two of them set our immortal souls a bit too unceremoniously to one side, the philosophy of particles in motion soon gained a reputation for atheism. But the popularisation of Father Gassendi's variant pacified pious souls and even became quite popular. More than that, some unique natural philosophical blends were produced in England. Merged with generous dashes of Stoic *pneuma* and with Plato's world soul, Descartes' subtle particles of matter were ground into an extremely fine, all-pervasive substance. In principle, that substance was material, but if needed it could also become the bearer of all kinds of spiritual effects (manifestations of ghosts) or even magically tinged ones. As a rule it was referred to as 'aether'. A procession of armchair philosophers touted it in all kinds of variants. And it did not stop there. In the 1660s–70s the aether was in its turn paired off with exploratory, fact-finding experiment. And thus was created what I shall call here the Baconian Brew.

The pairing off occurred in the work of the three greatest English exponents of experimental research: Boyle, Hooke and the young Newton. All three aimed at curbing the arbitrary assumptions that so often marked experimental research and the even greater fanciful theorising in particle thinking. A partial fusion, whereby one approach held the other in check, could provide the needed discipline. In two sets of seven points, Robert Boyle set out very precisely and, for him, concisely how mutual curbing would operate:

Of the Use of Experiments in Speculative Philosophy

1. To supply and rectify our senses
2. To suggest Hypotheses both more general and particular
3. To illustrate Explications
4. To determine doubts
5. To confirm truths
6. To confute errors
7. To hint luciferous [i.e. 'enlightening', a very Baconian term] inquiries and experiments and contribute to the making them skilfully

Of the Use of Speculative Philosophy to Experiments

1. To devise philosophical experiments which depend only, and mainly, upon Principles, notions, and Ratiocinations
2. To devise instruments both mechanical and others to make inquiries and tryalls with
3. To vary and otherwise to improve known experiments
4. To help a man to make estimates of what is physically possible and practicable
5. To foretell the events of untried experiments
6. To ascertain the limits and causes of doubtful and seemingly indefinite experiments
7. To determine accurately the circumstances, and proportions, as weight, measures and duration etc. of experiments

Here too, pairing off involved the transformation of particle thinking from an all-inclusive dogma into a source of hypotheses and other helpful tools. Particle philosophers of the current 'Athenian' style tended to fuss about whether the world is a void space through which atoms speed along (Gassendi) or is filled to the brim with vortices (Descartes). In typical 'Westphalian' spirit, Boyle felt that emphasising what both variants of the doctrine of moving particles had in common was much more fruitful than endless bickering over their differences. He reduced the several variants to what he called the 'catholick principles of matter and motion'. And the domain in which he set out to find out how fruitful these principles were was chemistry.

In this undertaking Boyle followed van Helmont quite closely. Except that all those effects that the vitalist van Helmont attributed to certain life-giving seeds were attributed by Boyle to moving particles of matter. In Boyle's view, the smallest corpuscles cluster together into relatively stable, larger units, 'primary concretions' (here the reader may think, if only for a second, of molecules). In many chemical reactions that come about by combustion or distillation or some other treatment, various combinations occur without anything essentially new taking place. You can, for example, put silver or mercury through a whole series of reactions in which substances of various kinds alternately manifest themselves, and finally recover the original silver or mercury unscathed. In such a case, there has merely been a reconfiguration of what already existed; the primary concretions remain unchanged. But according to Boyle it may also happen with some regularity that the primary concretions fall apart in a chemical reaction. Such an event allows the smaller particles that constitute them to enter into wholly new combinations and thus form new primary concretions. Hence,

> by the intervention of some very small addition or subtraction of matter, (which yet in most cases will scarce be needed,) and of an orderly series of alterations, disposing by degrees the matter to be transmuted, almost of any thing, may at length be made any thing.

Ultimately, according to Boyle, there are no bounds to the plasticity of matter. Consequently, many of his experiments were directed towards distinguishing between those reactions in which only a rearrangement takes place and those in which much wilder transmutations occur. One can already sense it: Boyle was practising alchemy just as enthusiastically as many others at the time, albeit within a much more thoroughly developed and much more profoundly thought-out theoretical framework. But it was not only alchemy that lent itself to far-reaching transformations. Boyle naturally realised that water cannot be converted directly into oil and certainly not into fire. But if you kept on giving water to a particular plant and distilled out its sap, then in his experience you could extract such large quantities of oil and carbon (which is, in a way, solidified fire) that it must have come from the copious quantities of water, not from the small amount of plant fibre.

In the eyes of posterity Boyle's systematic research into the less exciting of the two kinds of reconfiguration produced much more solid results than the one that, to him, counted most. If the treatment of one substance containing mercury turned out to produce a red powder and the treatment of a different substance produced a powder that looked exactly the same, then unlike many of his predecessors Boyle would undertake tests to see if by any chance it might be the self-same mercury compound. He was not the first to carry out identification tests on chemical substances, but he was the first to do it systematically. At this level, the pairing off of his corpuscular hypothesis and his experimental work proved to be fruitful indeed.

In a somewhat different way, Hooke's work also looks at first sight like a confusing mix of the bizarre and the fruitful. Hooke was not a strict or consistent thinker and doer; his was the clever, at times brilliant insight rather than any sustained elaboration thereof. Time and again he had a sense, which he certainly did not keep to himself, that others, Huygens and Newton among them, had made off with his discoveries. In so doing he underestimated by the tiniest little bit the difference between a flash of insight and a theory that has been explored to its very depths, or between a single fleeting experiment and a systematic series of them conducted with meticulous precision.

The difference would sour the last twenty years of his life when he fell out with Newton for good. But for a 'Curator of Experiments' his approach was ideal; every Wednesday he could instructively liven up the sessions of the Royal Society and help introduce some degree of coherence to the learned discussions without their being pressed into the service of some obsessive problem of a Curator's making.

Hooke's strategy for introducing greater discipline into corpuscular thinking was to exploit analogy more deeply. Let us recall the meaning of 'analogy' in this context. The hypothesised particles were invisibly small and their movements could not be observed even under the microscope. How then could one prevent the making of wild or even wilder assumptions about some explanatory particle mechanism? As well as the criteria of 'indisputably certain foundations', 'consistency' and 'physical intuition', a further criterion appeared to reside in the drawing of analogies with our visible macro-world. Together with that very informal 'intuition', it was actually the only available criterion at the experiential level. In Hooke's view, the essential analogy is provided by vibration. Particles vibrate incessantly. If their vibrations are in harmony with each other they blend well or even cohere into a single object, whereas disharmony keeps them apart:

> Those [particles] that are of a like bigness, and figure, and matter, will hold, or dance together, and those which are of a differing kind will be thrust or shov'd out from between them; for particles that are all similar, will, like so many equal musical strings equally stretcht, vibrate together in a kind of Harmony or unison.

Thus consonance and dissonance are turned into key elements of corpuscular thinking. Consequently, experimental research into musical phenomena gains the status of research into the basic structure of the material world. The analogy is even closer. According to Hooke, the 'like bigness, and figure, and matter' of the particles correspond respectively to the thickness, tension and length of a string in the macro-world. Mersenne had shown experimentally that these three qualities of a string determine the pitch of the sound it produces. Hence Hooke's fondness for experiments with sound, especially musical sound. He went beyond carefully testing Mersenne's results and attempted to produce consonant intervals by means of brass cogwheels. He also carefully recorded the shapes formed by sand or flour on the bottom of a glass bowl which had been made to vibrate in different ways (these have become known as Chladni figures after their late eighteenth-century rediscoverer). In the invention and execution of a great variety of experiments of this type, Hooke was a master.

The aether that Hooke put into permanent vibration in this manner became the bearer of a wide range of natural phenomena ranging from light to heaviness. For each of them, Hooke had an appropriate vibrating mechanism on offer; his powers of imagination were not that inferior to those of Descartes. But the closer his vibrational explanations came to living phenomena, the more he ran into difficulties. His pioneering work with the microscope confronted him with phenomena such as fermentation and rising sap in plants. How could they be explained by nothing but the vibration of particles? Indeed, where did those vibrations themselves actually come from?

In feeling his way towards answering this kind of question, the ambiguous character of the aether became clearly apparent. The aether was a conglomeration of vibrating particles, but at the same time it was, as it were, perfused by what Hooke called 'active principles'. Not every activity in nature can be explained by dead particles in motion; there is more going on. But can that 'more' be reduced in the end to some mechanism of dead particles? This was in due time to become the key question, and it is impossible to catch Hooke in the act of taking a consistent attitude in regard of it. At times he saw natural events as entirely material, but at other times without batting an eyelid he could produce such heresies as: Matter and Motion are the two fundamental principles, of which the first is the 'Female or Mother Principle ... [which is] without Life ... [It is] a Power in itself wholly unactive, until it be, as it were, impregnated by the second

Principle', that is, motion. And that second principle he went on to name 'Spiritus', the Latin term for Stoic *pneuma*. Such language harks back beyond Descartes to van Helmont and Paracelsus who put so much spirit and magic into their explanations. That is what you get when, on the one hand, you work with a rich aether in which *pneuma* and the world soul are battling with subtle matter for precedence while, on the other, your explanations of nature do not step back before the living world in all its spontaneous activity. Doing both these things simultaneously places corpuscular thinking under huge stress, and in Hooke's treatment of it we can see him running headlong into its outer boundaries. Take but one more step and the elastic band, now stretched to its very limit, is bound to snap.

A younger contemporary and compatriot inherited these ambiguities, and because he was a much more stringent and consistent thinker, it was in his hands that it did finally snap. We are already acquainted with that contemporary, who was seven years younger; it was Isaac Newton.

We left him in his 25th year. In 1668 he realised that he was becoming trapped in insoluble contradictions in his efforts to draw up a coherent account of motion which would not only, in unknown parallel with Huygens, harmonise the approaches of 'Athens-plus' and 'Alexandria-plus', but which in addition would be based on a mathematically defined concept of force of impact. In 1668 he called it a day and turned his attention to the Trinity, on which he rapidly developed deeply heretical thoughts, and to chemistry and alchemy. It was alchemy which eventually took him beyond what even with 'active principles' could still be explained within a sustained, corpuscular framework.

From the very beginning of his immersion in nature-knowledge Newton displayed an almost perverse penchant for studying precisely the kind of phenomena which the prevailing doctrine of particles in motion was least equipped to deal with. Take the ample capacity of air to expand, as in an air pump or indeed in many more common situations. An explanation that involves only some bare particles and their movements is clearly unsatisfactory; the situation almost screams out for a principle of repulsion. Newton's systematic study of the alchemical literature and the hundreds of experiments which he carried out with his own hands strengthened his conviction that even the 'vegetative processes' in nature cannot be reduced to mere movements of particles. In 1669 he wrote, still only for his private use, a summary with the revealing title 'On the Vegetation of Metals'. There it is clear that he, like Hooke before him, had long ago passed beyond the limits of what could still be fitted into some orthodox framework of particles in motion. 'Mechanicall coalitions or seperations of particles' are not sufficient to explain natural activity; 'wee must have recourse to som further cause'.

For the time being he sought that cause in an aether which was no longer merely Descartes' 'subtle matter', but was now indissolubly blended with *pneuma* and the world soul. Newton felt sure that the existence of the aether is demonstrated experimentally by the phenomenon that the oscillations of a pendulum die away in a vacuum at almost the same rate as in the air. The considerable resistance which the pendulum meets in the void must therefore come from the aether.

Newton not only adopted Hooke's universal vibrations for his aether but also fitted it out with locally varying density. It thus turned into the richest, the most abundant aether ever, and its explanatory powers were accordingly impressive. Everything that he had discovered about light and colour during his wonder years (with which I shall conclude this chapter) flowed smoothly from the local rarefaction and condensation of the aether. Phenomena such as heaviness or rubbed amber attracting snippets of paper could be explained by a kind of aether shower. Newton even extended his speculations over the entire natural world. 'Perhaps the whole frame of Nature may be nothing but aether condensed by a fermental principle', he wrote with the same ambiguity that Hooke was busily becoming entangled in. For was that 'fermental principle' an active principle within the framework of moving particles, or did it fall outside it?

In 1679 the elastic snapped. His continuing speculations about the aether persuaded Newton that along this route ambiguity could not be avoided. He even began to suspect more and more that, within the fine structure of matter which in his chemical and alchemical experiments he could almost grasp with his hands, there were *forces* at work. And from his earlier work on colliding bodies he still remembered very well that forces may be subjected mathematically to measure and rule in a way quite unthinkable with complex aether mechanisms and vague 'active principles'.

We do not know precisely how and when Newton made the leap from the micro-forces which he had begun to perceive in the vegetation of metals to replacing all active principles by the action of forces. What is certain is that in that same period he revisited his experimental proof of the existence of the aether. He reasoned that, in damping down the swing of a pendulum, the air can operate only on the outside of the pendulum whereas aether in addition penetrates the pendulum's pores and thus offers resistance from within as well. With extreme care he prepared a sophisticated experiment from which he inferred after doing his calculations that no resistance whatsoever is exerted on the inside of a swinging metal box. There was therefore no evidence that any aether was present.

With their respective aethers, Hooke and Newton had both gone far beyond the limits of 'ordinary' particle thinking on the lines of Descartes or Gassendi. But, unlike Hooke, Newton had now worked his way to the outer boundaries of what even the Baconian Brew, that blend of exploratory experiment and aether speculation, was able to contain. Up to its limits or already beyond them? The revised pendulum experiment persuaded him that there is no aether on or around the Earth. So it just might be the case that the effects that Newton had always ascribed to the aether were really due to forces operating far more widely than the micro-forces evidently active in chemical and alchemical reactions. Having come so far, he was able to react with equanimity to the challenge which faced him in 1679 due to a brief correspondence with Hooke.

The subject was orbital motion. Thirteen years earlier Newton had busied himself with the subject in connection with the fall of apples and the orbit of the Moon. He had begun to suspect that in a sense the Moon is constantly falling towards the Earth but that there is something that prevents it and forces it onto a trajectory around the Earth. Or rather, at that time Newton still saw it the other way round, in accordance with the basic concepts of particle thinking: an object, or a heavenly body, is on the one hand under pressure from particles that hurl it round in a circle, while on the other hand at every point along that path it tends to fly off in a straight line. Hooke now confronted Newton with the opposite view on orbital motion: an object, or a heavenly body, moves uniformly in a straight line but is continuously pulled off that line by an attractive force. Hooke was thinking of some kind of magnetic effect, and this helped him suspect in addition that the force diminishes in proportion to the square of the distance. In this specific case, the motion in a straight line will be bent into an ellipse, as in Kepler's planetary orbits.

Hooke lacked the mathematical skills needed to subject such a suspicion to rigorous proof. But Newton did know how to do it and all the more so because he had already had a similar suspicion much earlier, though at the time the results did not seem to tally. Without involving the despised Hooke further, Newton wrote the proof down for himself. *Mathematically* it was a masterpiece – nobody else would have known how to make the transition from a force that acts in minute yet discrete portions to a continuous force. *Conceptually* he now had in his hands, thanks to Hooke, the innovative perspective needed to work out in systematic fashion his budding thoughts about forces. Only, his mind was on other things, like alchemy and theology, so he put his proof away in the same drawer where so much more of his work already lay waiting to be completed.

Seven years earlier Newton had been elected a Fellow of the Royal Society, but he disliked society, which always entailed the risk of encountering people who might disagree with his views. So he stayed away from the noisy discussions on Wednesdays and in the London coffee houses (coffee was then, together with tobacco, the drug in fashion). Hooke in contrast was thoroughly in his element. If he was not at home experimenting, he could be found in a coffee house exchanging news and bandying ideas about. The problem of orbital motion was one of dozens of subjects which were regularly discussed, in particular between him and a group of mathematically inclined Fellows who included the astronomer Edmond Halley. Hooke claimed that he had proof for his assumption that an attractive force which diminishes in proportion to the square of the distance will make a heavenly body that travels uniformly in a straight line deviate onto an elliptical path. Only, he would rather not reveal that proof until it was apparent that nobody else was able to produce it. This was a typical Hooke gambit, and he would live to regret it. For Halley was keen to know whether Hooke's claim was correct and when, in the summer of 1684, he had to spend time in Cambridge he decided by way of a side visit to knock at the door of Professor Newton in Trinity College. Rumour had it that as a mathematician he was up for a challenge, so perhaps he would have some thoughts on the nature of orbital motion.

THE GREAT SYNTHESIS: NEWTON ROUNDS OFF THE REVOLUTION

His host most certainly did have some thoughts about it. More than that, he had proved the elliptical orbit years ago, after his brief correspondence with Hooke. Halley was mathematician enough to realise that this raised the issue to a whole new level. 'Struck with joy & amazement', he eagerly begged to see that proof. But Newton claimed that he could not find it and promised to send it to Halley later. Three months later Halley received the proof, but not the proof alone. It now formed part of a succinct treatise. In nine pages, Newton laid the mathematical foundations for nothing less than a general theory of the action of forces, plus a still vague hint of an attractive force that might hold the planets in their orbits. Halley appreciated its huge potential, sprang into the saddle and went back to Cambridge, this time with the single objective of persuading Newton to develop the nine pages more fully. There was now no stopping Newton, and two and a half years of furious work ensued. As he advanced, Newton was gripped by that delightful feeling that can overwhelm a person who finds that he or she has now hit upon the conceptual framework in which every new discovery just leads on to another, and another. Not only did a rapidly growing number of phenomena appear to fit, they even appeared to hold up against all empirical checks. He hardly slept, he refused to be distracted by anyone or anything, he ate whenever it happened to be convenient and sometimes not even then. He might get an idea on the way to high table, return to his room and work until deep into the night. Among the Fellows of Trinity College, whose thinking had not even made it as far as Descartes, let alone left him behind, one anecdote after another piled up about their bizarre colleague and his bizarre obsession.

The formidable intellectual strides that Newton took in the two and a half years spent composing the *Principia* amply deserve to be followed step by step. Here we must pass them over, and confine ourselves to the actual contents of the book and its apparent implications for the future pursuit of nature-knowledge.

The full title is *Philosophiae Naturalis Principia Mathematica*, 'Mathematical Principles of Natural Philosophy', or less literally, '... of Natural Science'. For the book had nothing left in common with natural philosophy in the 'Athenian' sense. So for us the time has come to exchange our historical concept of 'nature-knowledge' for the modern term 'science'. Surely, the mathematical style of expression in the *Principia*, then brand-new, is now obsolete. Also, in the second edition Newton introduced God in Person (One Person not Three). But for the rest, a present-day physicist can read and understand the book as if it were written by a respected former colleague.

It is no accident that the title alludes to Descartes' *Principia philosophiae*. The message is clear: if you want to lay *proper* foundations for natural science, then you can only do it mathematic-ally, not verbally as Descartes had done. And so the book opens with the general laws of motion with which Newton replaces those of Descartes.

The first law is our old acquaintance, Galileo's and later also Beeckman's and Descartes's idea that, without outside interference, motion will be retained forever. What had remained ambiguous until then was now quite consistently pronounced by Newton to be valid for uniform motion in a straight line, and endowed forever with the name 'principle of inertia'.

The second law, in contrast, is a radical innovation. It states that a force does not bring about velocity but acceleration. 'Acceler-ation' not only in the sense of increasing or decreasing velocity, but also of change of direction. Uniformly accelerated motion in a straight line is not therefore the opposite of uniform angular motion in a circle. From the viewpoint of the force that brings them about, they both amount to the same thing. This insight, as paradoxical as it is grandiose, enabled Newton to treat mathematically every alteration of an inertial state, not just the one that is uniform in a straight line. The way was free to find out what paths would result from a huge variety of forces.

Newton paid especially close attention to the action of one particular force, the one by which an object is bent off its uniformly traversed path in a straight line by a force which diminishes in proportion to the square of its distance. At first, Newton derived the resulting elliptical path from an abstract kind of force without (at this point in his book) giving it any physical significance. But further on in the *Principia*, when he comes to 'The System of the World', he does precisely that. Its physical significance consists in the attraction which each particle of matter exerts on every other particle of matter. Inside an object, from an apple pip to a planet to the Sun, the attractive force must be treated as if it is concentrated at the centre. The larger the mass of the attracting object the greater the force of attraction, since it consists of a greater number of particles of matter. 'Heaviness' is a manifestation of that attraction: a body falls because it is being attracted by a body with a much greater mass. In the solar system, the planets rotate around the Sun because it attracts them, always in that fixed proportion of diminishing with the square of the distance. Hence, Mercury is attracted more strongly than Saturn – Kepler's third law is an immediate consequence. Naturally, the attraction is mutual. The apple also attracts the Earth even though it is not noticeable because their masses differ so enormously. But sometimes the differences are smaller. The planets for instance all attract each other; between Jupiter and Saturn the resulting deviation from pure ellipticity could even be measured. His pursuit of such effects confirmed Newton in his conviction that he was on the right track with his universal gravitational attraction, which in any case had taken him a great deal further than any aether action had ever done.

Indeed, such disturbances of planetary orbits appeared to be confirmed by the observations which the best astronomers of the time had either published or sent to him when specifically asked. Tidal ebb and flow provided another pointer; the tides appeared to lend themselves rather well to being explained by the combined attraction exerted by the Sun and the Moon on the Earth's oceans. And above all it appeared that the data upon which he had, twenty years earlier, established a correlation between the fall of an apple and the orbit of the Moon that he had by then laid aside as not sufficiently precise could be replaced by more recent and better data which now showed the correlation to be valid with complete exactitude.

Meanwhile it behoved Newton to investigate whether any alternative force actions existed that might lead to the same or even better results. In theological terms, the question was whether God in creating the universe had had any choice in the matter. Could He have employed different forces, for instance one that diminishes in proportion to the cube of the distance, or one that is proportional to the distance? Hence the thorough survey that Newton made in the *Principia* of all those other forces in the abstract – what orbits would they produce and how stable are they? Or those vortices of Descartes: are they not also capable of producing a stable universe that looks the same as ours? In order to answer that last question in particular, more than a little pioneering research was needed into the action of forces in a medium that offers resistance (as with motion in air or in liquids). The conclusion at which Newton arrived in the end was that God, if He wanted to create a viable universe, had no choice – it cannot be done without the inverse square law. So, in addition to all His other excellent qualities, God was also a first-rate mathematician who as far back as the Creation had gone to work with great care in this sense, too – a formidable sign of intelligent and also most careful design!

Newton undertook all these investigations into a multiplicity of possible force-actions-in-the-abstract for yet another reason. He had meanwhile become convinced that our world is full of forces of the most diverse kinds. Not only the universal force of attraction which he had now discovered, but also other forces, exert an effect such as in chemical reactions or in electrical attraction. Only, the action of such forces upon such phenomena had yet to be discovered. In the *Principia* Newton subjected in advance the general properties-in-the-abstract of a whole range of possible forces to preparatory mathematical investigation.

What enabled Newton to write this truly pioneering book? The question can be answered on more than one level. At the personal level, there were the salutary interventions of Hooke and Halley. Unintentionally, Hooke had helped Newton to reformulate his initial conjecture of a correlation between the fall of an apple and the orbit of the Moon in such a way that he could press on further. And Halley was not only the man who with his two visits persuaded Newton to press on further indeed. He was also instrumental, practically and even financially, in ensuring the publication of the book. Without that letter and that first visit, Newton might never have embarked upon the *Principia*, let alone completed it. Yet neither intervention was just a happy coincidence. At the institutional level the Royal Society was involved in all kinds of ways. It was as secretary of the Society that Hooke wrote to Newton in 1679. Halley served at the time as its clerk. And the problem of orbital motion was discussed by the Fellows with some frequency during their sessions and in the coffee houses. Newton further owed his up-to-date data to the journals and the correspondence which gave scientific life of his time so much zest and momentum. But, in the end, what enabled Newton to bring about this sixth revolutionary transformation was situated at the level of ideas, concepts and theories. This was the unique circumstance that at an earlier stage he had also taken part in *both* transformations no. 4 *and* no. 5.

With no. 4 the pioneer had been Huygens, his senior by thirteen years. Why was it not *he* who discovered the second law of motion and universal gravitation? It is very characteristic that, in a sense, Huygens did discover the second law of motion – that force produces acceleration – independently of Newton. Around 1675, hence twelve years before the *Principia*, he wrote for his own use a short note in which he coined a new concept of force. He called it 'incitation' and understood it to be that which brings about acceleration. He gave a few examples and then terminated the entry, for good. Why did he never take the idea any further? Because his thinking and that of his immediate circle lacked precisely what marked the Baconian Brew on the other side of the Channel – a rich aether with all its ambiguities. Newton had to work his way experimentally and speculatively all the way through the aether and its active principles before arriving, in the end, at forces. Huygens always remained faithful to his unambiguously straightforward particles in motion.

Before Newton, such a rich aether had been explored by Hooke, his senior by seven years. He was the man who even in 1679 could still teach Newton a thing or two about orbital motion, and who had himself reached the limit of what 'active principles' could still achieve in particle thinking. Why, then, did *Hooke* not discover the second law of motion and universal gravitation (not that Hooke himself saw it in that way)? Just so, but now in reverse, Hooke had played no part in the fourth transformation. Put another way, Hooke lacked the mathematical training and in particular the mathematical rigour and discipline to give his conjectures about aether a firm foundation and a solid form.

In short, to find the law of universal gravitation it was not enough to be either Hooke or Huygens. Only a thoroughly blended 'Hookgens' could come up with something like that. And no single person met that requirement but Newton alone.

A tricky question that arose from Newton's introduction of all those forces, in particular that of attraction, was whether he was not in effect reverting to the hoary idea of occult forces. For Huygens and his former pupil Leibniz, the most prominent critics of the *Principia*, the answer was a definite yes. They were convinced that Newton's forces could pass muster only if he traced them back in their turn to the action of material particles. Meanwhile Hooke, for whom this was less of an issue, was telling everyone in London who cared to listen that it was a disgrace, and that Newton had stolen universal attraction from him.

Newton's response to both sets of criticism was typical. To Huygens and Leibniz he replied that he too did not know the precise nature of those forces, but this did not make them occult in the usual sense of the word. This was so because he had now demonstrated it to be possible (as it was not with occult forces) to define exactly, that is, mathematically, the action of the forces that he had introduced, in his laws of motion and all that ensued from them. Furthermore, there were numerous phenomena, from the tides to comets and how Jupiter and Saturn affect each other's orbits, which sufficiently demonstrated the exactness of their action. And to Hooke he replied in the most offensive terms that he seemed quite oblivious to the difference between *asserting* a thing and *proving* it.

Indeed, in the *Principia* Newton had furnished proof, in great detail and with all the rigour then possible, that a body moving uniformly in a straight line and subjected to the action of a force directed at the centre and diminishing with the square of the distance will describe an ellipse. Hooke had suspected it; Newton had initially come up with a quite succinct proof, and had now reworked his proof much more thoroughly and exhaustively. This time he produced the required proof from two new, immensely fruitful concepts of force that he had thought through to the limit and explored to many a remote corner, the one the physical embodiment of the abstract other. And he had done so without skipping a single step along the way. The message to Hooke and the world at large was clear: throwing up an idea is one thing, firmly nailing it down mathematically and experimentally is quite another.

In the *Principia*, then, Newton deliberately laid down the new criterion so urgently needed in his view for imposing the most rigorous limits possible on fanciful theorising in natural research. The criterion was not really new in the strict sense of the word. In essence we are dealing here with the same kind of balancing act between mathematical deduction and experimental testing that was explored in the footsteps of Galileo by the relative few who in the second generation fell under the spell of Alexandria-plus. Newton, of course, was one of them. Only, in the *Principia* he gave greater substance to the balance through the tight coupling that he was able to make time and again between abstract deduction from a multiplicity of force actions and their application in the solar system. There were two main reasons for Newton to attach such paramount importance to the utmost exclusion of fancy and arbitrariness.

Firstly, his great immediate predecessors Huygens, Boyle and Hooke had retreated to the bastion of 'probability'. With more or less reluctance (the perennial hesitator Boyle more than the stricter Huygens) the three saw certainty in nature-knowledge as ultimately unattainable. With that they had in their turn set their face against the indubitable certainty which natural philosophers in the Athenian tradition had always used as a club to batter their critics. But for Newton, half a generation on, their step backwards from certainty to probability was a step too far. It was not necessary to make that concession, nor was it desirable.

It was particularly undesirable, and here lay the second reason, in that there was within Newton himself such a powerful tension between the certainty of mathematical deduction and the fancy-prone quest for a consistent conception of the world at large. He was not only a meticulous mathematician and rigorous experimenter determined never to say more than he could strictly ascertain to be factually true according to his own perfectionist standards. He was also addicted to speculation, to designing highly imaginative sketches of 'the frame of nature', whether composed of aether particles and a principle of fermentation or, as later, produced by the interplay of forces. His inability ever to resolve that inner tension provided a strong driving force for his work, which took shape in brief bursts of explosive creativity. In his publications, Newton virtually confined himself to the two books in which he felt reasonably satisfied that he had managed to maintain his own rigorous standards. But throughout his life he vacillated whether or to what extent he should also make his speculations public. In finding half-hearted solutions for this perennial dilemma he was a master. Not until the 1970s, when his posthumous notes began to be subjected to thorough-going historical analysis, did the world of scholarship become aware of Newton's successive world-views and their partially alchemical genesis.

I just spoke of 'two books'. The *Principia* appeared in 1687; the *Opticks* seventeen years later. Yet in many respects the second book is really the first; except for a few sections he could already have published it in or around 1672. In this book too, apart from at the end (where in question form he allows something of his broad world-view to filter through), Newton tried to maintain the most rigorous standards of proof possible. But here he succeeded less well, or rather he succeeded experimentally better than mathematically. Even so, thanks to his sustained and precise measurements he managed with his experiments on light and especially colour to attain a level of exactness that he hoped would set new standards, too.

Just as before and with the *Principia*, his great rivals were Huygens and Hooke. Again the rivalry was partly simultaneous and partly unwitting; again it gave rise to polemics. Let us leave aside all the events of the 1670s and confine ourselves to the key question: in comparison with the work of the somewhat older twosome, what makes the *Opticks* special as a contribution to curbing 'fancy', that arbitrary element in natural research?

Opticks is about light and colour. It was not so long previously that those subjects were brought together for the first time. Before the transformation of 'Athens' into 'Athens-plus', explanations of light were closely linked to vision; in particular Ibn al-Haytham (Alhazen) had bound them tightly together. Colour was treated as a quality of objects, not of light. But there are also colours without an obvious object, such as the rainbow. In that case, so the reasoning went, white sunlight under the influence of the intermediate matter (air or water) undergoes a modification in which it adopts all the colours. Descartes saw light as the propagation through space of the pressure which particles of matter exercise on one another as each incessantly pushes all neighbouring particles out of its way. According to him, colour was connected with the axial rotation of the particles: the quickest we observe as red, the slowest as blue and the rest in between. In this way colour became a property of light. But with all that, colour remained a modification, a metamorphosis of the original white light.

Meanwhile, Descartes, with his 'Alexandrian' hat on, had discovered the sine rule of refraction, which expresses how a ray of light is broken at the interface between two materials, for instance air and water or air and glass. For Huygens the sine rule was an indispensable tool in his research into the optimal combination of lenses for telescopes. But while he was thus occupied, in the 1660s two exceptions to the sine rule made their appearance. A particular kind of crystal, called Iceland spar, refracts a ray of light in two ways at the same time, and one of them deviates considerably from the sine rule. The discovery initially left its students baffled, but it inspired Huygens to one of his finest discoveries. In 1679 he gave a series of lectures at the Academy which he published eleven years later under the title *Traité de la lumière* ('Treatise on Light'). In it he defined light as a succession of pulses transmitted by particles of matter. This happens in accordance with his laws of impact, thus in a straight line, as is appropriate for light. As the pulses are transmitted along, a wave-front is produced that is perpendicular to the direction in which the light is

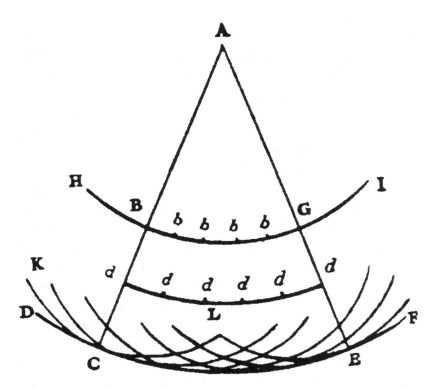

Huygens' principle
ACE is a beam of light. The points b and d on the wave-front that extends from A each form the centre of small wave-fronts. The wave-front DF is made up of all the small wave-fronts.

travelling. Huygens' big find, still known as 'Huygens' principle', was that this comes about because every affected particle begins to function in its turn as the source of a tiny wave-front.

This supposition now made it possible for Huygens to derive mathematically not only the ordinary phenomena of light, reflection and refraction, but also, in particular, the strange refraction in Iceland spar. As with his work on collision and on other types of motion, his explanation of light fits into the fourth revolutionary transformation in which problems thrown up by Descartes in natural philosophy were resolved in the realist-mathematical style of Galileo.

Meanwhile another, more profound instance occurred in which the sine rule of refraction fails to apply. One of Newton's great discoveries during his 'wonder years' around 1666 involved colour. If you direct a beam of light at a glass prism, and set up a screen some distance away, the light after refraction does not continue as before but appears to split into separate colours. What you see on the screen is an elongated band with blue at one end and red at the other, in other words a spectrum in which the degree or index of refraction is slightly different for each colour. It is not white light that is primary, but the colours, and white light is a combination of them.

This discovery deviated radically from what everybody had always taken to be self-evidently the case. When Newton published it in 1672 in the journal of the Royal Society, the *Philosophical Transactions*, both Huygens and Hooke initially missed this central point. What caught their eye was the suggestion that the phenomenon of strange refraction in prisms had to do with the particulate nature of light. Throughout his life, Newton remained convinced that light is not a pulse or wave, but consists of rapidly emitted particles. The laborious polemics into which his critics drew him as a consequence of this misunderstanding contributed to his determination henceforth to keep speculative explanation and hard facts and proof entirely separate. What did he actually have in the way of hard facts and evidence?

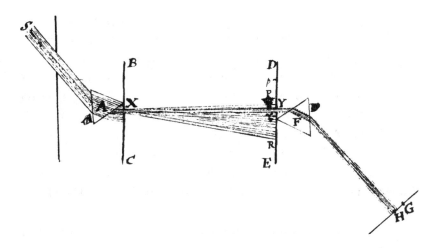

Newton's crucial experiment
Light from the Sun, S, shone through a hole in the wall of Newton's
student room onto a prism A. On screen BC appeared a spectrum. Light of
a particular colour passed through hole X in the screen. Newton could
vary the colour by rotating the prism. Through hole Y in the second
screen DE rays of that colour passed through a second prism F. At screen
GH it appeared that homogeneous light of whatever colour undergoes no
further shift. In short, 'ordinary sunlight is a heterogeneous mixture of
rays, each with its own immutable degree of refrangibility'.

He had much to prove indeed. The elongated shape of the
spectrum, like the colours themselves, could surely be another
modification of white light, artificially induced by the prism? In his
famous *experimentum crucis* (crucial experiment) Newton showed
that if, after the first colour sifting, you separate light rays of one
specific colour and let them pass through a second prism, no further
shift or elongation takes place. Another powerful experiment
showed that when, after refraction, the whole spectrum is then
concentrated again on one single point, the colours immediately
merge to form white.

They were ingenious demonstrations, which sooner or later
persuaded most readers. But the decisive advantage of his discovery
was that it lent itself so well to precise measurements. In making

them he did not limit himself to prisms; there were other ways of separating out colours. Hooke had discovered one of these ways with his microscope: in thin layers of mica, rings are formed which display all the colours of the rainbow. However, the rings did not lend themselves easily to direct measurement, so Hooke, unable to do more, made do with one of his quick explanatory hypotheses. But in Paris and Cambridge there were two mathematical scientists who, on reading the passage in Hooke's book, realised at once that considerably more could be made of it. One of them was the recognised grand master of natural research; the other had just graduated. Both men saw straight away what mathematical tools were needed to make the measurements, indirectly yet with so much the greater precision. In their measuring, both stumbled on the periodic nature of the phenomenon: it always repeats itself at the same distance. In their measuring, both stumbled on the same side effects and the same apparent anomalies. Confronted by these anomalies, Huygens soon gave up on them, while Newton persisted:

> Newton's skill in performance ... outran that of Huygens from the beginning, and in the more sophisticated experiments of 1670 he simply eclipsed his unsuspected rival. What he demanded of his measurements tells us much about the man. Measuring with a compass and the unaided naked eye, he expected accuracy of less than one-hundredth of an inch. With no apparent hesitation, he recorded one circle at 23 ½ hundredths in diameter and the next at 34 ⅓. When a small divergence appeared in his results, he refused to ignore it but stalked it relentlessly until he found that the two faces of his lens differed in curvature. The difference corresponded to a measurement of less than one-hundredth of an inch in the diameter of the inner circle and about two-hundredths in the diameter of the sixth. 'Yet many times they imposed upon mee', he added grimly to his successful elimination of the error. No one else in the seventeenth century would have paused for an error twice that size.

It is true that in these investigations more was at stake for Newton than for Huygens, in whose optical research colour did not play any significant role. As was most often the case with Huygens, he had hit upon a fellow investigator's discovery and perceived, better than his peers, how to improve on it. But the foregoing does teach us something about the contrasting demands that the two of them made of a truly scientific approach – Huygens' idea that more than probability is not achievable in any case does not in the end advance the remorseless hunt for precision of the kind that singled out Newton.

For Newton, precision was indispensable in the search for truly adequate explanations. He had no time whatever for the kind of hypothesis that Hooke could pluck out of the air with such appalling ease. In the correspondence that ensued when Newton offered his discovery of colour sifting for publication to the first secretary of the Royal Society, Henry Oldenburg, this young mathematics professor who was still quite unknown outside Cambridge wrote down his *credo* on the subject:

> The Theory wch I propounded was evinced to me, not by inferring tis thus because not otherwise, that is not by deducing it onely from confutation of contrary suppositions, but by deriving it from Experiments concluding positively & directly.

In his initial letter Newton had already written about colour dispersion that it is 'the oddest if not the most considerable detection wch hath hitherto beene made in the operations of Nature'. However boastful this may sound, fifteen years before the *Principia*, he was not exaggerating. Always and everywhere, everybody had taken it for granted that bright, white sunlight is the original light. It went against all intuition to suppose the opposite, to observe it and to persist in it.

Similarly, it had flown in the face of intuition to argue that the Earth is not the motionless centre of the universe but just one of the planets circling the Sun. More than half a century had elapsed between the half-hearted suggestion and the observational evidence for it, as it had taken half a century to eliminate Copernicus' half-heartedness in the first place. But with sunlight and colours all this happened in one throw. With its gradual acceptance at the end of the seventeenth century it had become clear that nature does not lend itself well to being studied with the obvious tools of common sense and the unassisted five senses. Nor is it susceptible to speculative deductions from first principles, and only to a limited extent to purely qualitative description. To fathom nature, one must be prepared to plumb the depths, learn the language of mathematics and accept and apply experiment as a source and touchstone of knowledge attained wherever possible by means of precision measurements. Remove mathematics and experiment from Newton's work and what remains are the visionary outbursts of a highly intelligent fantasist, not the mature, elaborate conclusions of the man who brought together and synthesised the widely diverse results of scarcely a century of revolutionary innovation in nature-knowledge.

Those were the guidelines for natural science which, less than a century after Kepler, Galileo, Descartes and Bacon and all the rest of them, lay encapsulated between the covers of Newton's two books. With those two books the Scientific Revolution came to an end. At least, it did so in a historical sense – *Principia* and *Opticks* mark the final stage of one clearly demarcated historical episode.

Epilogue: a look back and a look ahead

Does it matter whether we characterise what happened in seventeenth-century nature-knowledge as a revolution, or refrain from using that loaded term?

In one sense, of course not. 'Revolution' is just a word, with changing meanings which up to a point we may determine ourselves.

Nevertheless, we have in the course of this book come across several criteria that might help us to decide whether or not it makes good historical sense to keep speaking of 'The Scientific Revolution of the seventeenth century'.

A VERITABLE REVOLUTION?

Take for starters the vast contrast between the state of nature-knowledge in 1600 and that one century later. It is vast in terms of reining in 'fansy' which we discussed at length when exploring the different ways in which the seventeenth-century pioneers of a radically new conception of the natural world sought to confirm the validity of their conclusions without resorting, as in the past, to hoary dogmatic certainties or conversely lapsing into all-encompassing sceptical doubt. In realist-mathematical science, the interaction between mathematics and experiment appeared to offer the prospect of a progression of often valid, sometimes wrong, but as a rule well-testable conclusions. In exploratory-experimental science, numerous artful devices were tried out to pin nature down in all its capricious unpredictability; at the end of the century, Newton made the criteria for this even more demanding. In short, nature-knowledge in the seventeenth century served, as it were, as a laboratory for finding out how to probe specific truth claims (whether made by oneself or by anyone else) in a spirit of methodical distrust. It goes a little

256

too far perhaps to follow Popper here without more ado and ascribe 'falsifiability' to the pursuit of nature-knowledge in the seventeenth century, yet an implicit sense of human fallibility and a determination to make it operative (that is the new thing) are clearly taking over as the century proceeds.

The contrast between 1600 and 1700 is also vast in terms of substantive content. The Earth turns. Our blood circulates. The air that we breathe has weight. A void space can be created. Objects that are attracted by the Earth fall to the ground with uniform acceleration. White light is composed of all the colours of the rainbow. Unequal cross-sections of a river discharge equal quantities of water in equal periods of time. Human sperm consists of millions of tiny creatures. For the cycloidal trajectory of a pendulum the duration of an oscillation is independent of the deflection. The natural tones of a trumpet are linked up with multiple vibrations. The path of a bullet is a parabola. The digestive system neutralises the caustic effects of acid. Planetary orbits obey the area law. Electric repulsion exists. The list can be multiplied a hundredfold.

However little attention we may have paid in this book to developments in 'pure' mathematics, we have seen enough to notice there, too, an enormous contrast. At the start of the seventeenth century, essentially Greek geometry and some arithmetic provided the only quantitative tool for those who pursued nature-knowledge in Alexandrian style. Less than a century later, Newton's differential and integral calculus enabled him to get a grip on the plethora of natural phenomena that, in the *Principia*, he subjected to an investigation steeped in infinitesimal geometry and algebra.

Hardly less vast is the contrast in terms of society and its institutions. At the start of the seventeenth century, nature-knowledge was a marginal phenomenon and natural philosophy in particular was linked in various ways to theological dogma. A century later, natural science had attained an unprecedented degree of autonomy and visibility, centred on two major societies which operated on a grand scale and staged often highly productive, blow-by-blow exchanges (both oral and in writing) between unprecedentedly large numbers of practitioners.

This enormous, at the very least fourfold, gulf separating the beginning and end of the seventeenth century is far from the only criterion for 'revolutionary, yes or no' that we have encountered along the way. By definition, revolutions challenge the established order. No matter how, they reduce an 'ancien régime' to a thorough state of confusion even if the revolutionary process leads in the end to some form of restoration. How far is that true of what happened to the state of nature-knowledge in the course of the seventeenth century?

So much is certain that by the end of the Scientific Revolution little remained of the 'ancien régime' of pre-seventeenth-century nature-knowledge. Not that it surrendered without putting up fierce resistance against all those new-fangled knowledge claims – a resistance that was stimulated not only by centuries-old patterns of thought and theological-political persuasions and vested interests, but also by a genuine, quite understandable lack of comprehension of what the innovators were actually seeking to achieve. Whether we think of the able astronomer Peter Crüger puzzling in vain over the enigma of Kepler's bizarre 'celestial physics', or of Galileo's Pisan colleagues and their thorough-going Aristotelianism, or of Hobbes' unshakeable conviction that no vacuum was produced in Boyle's air pump because his own natural philosophy ruled out that possibility *a priori*, in each of these and countless similar instances to all intents and purposes one can speak of almost completely incommensurable mental worlds. In cases such as these it makes sense to speak of 'paradigm shifts', the hallmark of scientific revolutions in the sense which Thomas Kuhn has given to the term. Between 'Alexandria' and 'Alexandria-plus', between conceiving of mathematics as inappropriate or appropriate for grasping basic properties of the natural world, there was a yawning gulf of perception which, as with all genuine paradigm shifts, could be bridged only by gifted young minds such as those of Torricelli or Huygens. For the two other revolutionary transformations which took place between about 1600 and 1645 the situation was similar, if less pronounced. But above all else, we have seen that by the end of the seventeenth century the 'old' modes of nature-knowledge were effectively played out. Even the five subjects that had occupied practitioners of the mathematical mode of nature-knowledge during each flourishing period (planetary trajectories, light rays, consonant intervals, equilibrium states of solids and fluids) were now fully absorbed into science of the new realist-mathematical kind. Something similar is true of fact-finding experiment and its predecessor, accurate observation, while natural philosophy as a speculative-dogmatic mode of acquiring knowledge had by about 1700 landed on the back burner where it has remained ever since.

Another criterion rests in the emergence of two entirely new hybrids. Before 1600 we encountered the following mixed forms: (1) dogmatic, natural-philosophical pseudo-systems which blended doctrines from the various Athenian schools; (2) incidental use of some suitable doctrine borrowed from natural philosophy to plug occasional gaps in a mathematical argument (Ptolemy, Copernicus); and (3) the extension of Aristotelianism in the direction of magic (Fernel) or mathematics (Clavius). In the course of the seventeenth century they were all replaced by more durable and fruitful hybrid forms. In the case of Horrocks and Kepler's three laws we saw how a latent productive element in the work of one thinker could be taken out of its original context by another thinker and made to reveal its true merits in a different setting. Even more incisive was the transformation of the speculative-dogmatic philosophy of moving particles into a broad hypothesis which could be made fruitful by combining it with realist-mathematical science (Huygens, the young Newton) or with fact-finding experiment (Boyle, Hooke, the young Newton).

Yet another criterion lies in the questions that the pioneers inherited from their predecessors and the questions that they in turn passed on to their successors. We discussed this in the most detailed manner with Kepler and Galileo. Kepler inherited from Tycho the traditional question of which eccentric, uniformly traversed, circular trajectory could best 'save' the observational phenomena of the planet

Mars. To his successors he left the question of the interconnection between the really and truly elliptical path of not one but all the planets, the area law which determines their speed around the Sun, and the fixed proportion between orbital period and mean distance to the Sun that forms Kepler's third law, as well as the question of which specific forces underlay them all. Galileo inherited the age-old question of the cause of an object's initial acceleration in free fall and left behind the much more specific and now actually answerable question of why objects in free fall accelerate uniformly. Again much the same happened in natural philosophy and fact-finding experiment. Before Beeckman and Descartes, the only development in natural philosophy that went beyond mere recycling was the expansion of Aristotelian doctrine in the direction of mathematics (Clavius) or magic (Fernel), leaving unchallenged (but for the usual sceptical criticism) the speculative-dogmatic way of thinking as such. After Descartes the leading question in natural philosophy became what plausible-looking variations on the master's whirlpool mechanisms could be thought up, while for a few individuals (Huygens, Boyle, Hooke, the young Newton) the urgent question arose whether all such speculations were not the hopelessly sterile products of fantasy run rampant. Bacon faced the question whether hyper-accurate observation and description were being carried out methodically enough to reach a coherent insight into the natural order; after Bacon the question was how (bearing in mind the readily apparent, wily tricks played by nature in Bacon-like experimentation) valid knowledge could be extracted from entire series of experiments. Gilbert, Harvey and van Helmont inherited from their predecessors, all of them skilled in accurate description, some specific problem concerning, respectively, magnetic attraction, the blood stream in the human body, and the composition of mineral cures. The question that they left behind was which specific tools (artifices, instruments, interventions in the natural state of a given object) were most suitable for experimental research. This large-scale discrepancy between the problems which the pioneers took over and those that they left to their successors was expanded during the second and third generation of the Scientific Revolution to the literally endless 'problem → solution → new problem → new solution' dynamic which the example of Huygens' pendulum clock so clearly illustrated and which has been characteristic of scientific research and its unceasing advance ever since.

So, criteria enough, one might think, to establish unambiguously the typically revolutionary character of what happened in seventeenth-century nature-knowledge. And yet ... Against the tenability of the historical concept of 'The Scientific Revolution' it is regularly argued that history is a continuous process and that there are never any complete breaks with the past, not even in the seventeenth century. Furthermore, it is pointed out that a full century is a very long time for a revolution. Even if one incorporates the Napoleonic period into 'the French Revolution', no more than just over a quarter-century separates the fall of the Bastille in 1789 from the Battle of Waterloo in 1815. In this book I have met the two objections by breaking with a persistent habit of treating the Scientific Revolution as a single homogeneous event. On the contrary, I have broken it up into six closely interlinked episodes of *revolutionary transformation*. The longest of these six episodes lasted (due to the late appearance of Galileo's *Discorsi*) for forty-five years; all the others were at least a decade and a half shorter. More importantly, each of the episodes displays quite specific characteristics and its own – at times somewhat varying – degrees of continuity and rupture. Furthermore, for each I have established a specific set of causes which clearly set it apart from the others.

Meanwhile, the primary criterion for 'revolutionary', in the sense of a transformation of an unusually incisive and far-reaching range and significance, lies elsewhere, namely in the very phenomenon which I sought to point out by means of our imaginary trend-watcher of 1600. His extrapolation of long-existing trends and the glaring contrast between what he predicted on the basis of those trends and how things actually developed is a feature of both markers of a genuine revolution: the disruption of a long-standing pattern and the complete unpredictability of that disruption. According to the old pattern, nature-knowledge would flourish for a few centuries and culminate in a Golden Age, but then its high level used to drop off quite steeply, even though the decline might still be punctuated by some local, incidental high points. Those who start their account of the Scientific Revolution back in medieval or Renaissance Europe, and even those who begin their narrative in antiquity but do not apply a consistently comparative approach, are bound to miss that pattern. However, once we recognise it we can see that the radical disruption and its completely unpredictable replacement by something quite drastically different (in our case, first by the appearance of three virtually simultaneous revolutionary transformations followed shortly afterward by the unbroken expansion of scientific research to which nowadays we have become accustomed) makes it eminently sensible to hold on to the concept of 'The Scientific Revolution'. But not to hold on to it *unchanged*. Ongoing historical research has sufficiently demonstrated that earlier conceptualisations suffered from a number of important shortcomings; but the shortcomings of one particular conceptualisation need not entail the abolition of the concept itself. I therefore present this book as an effort at a radical reconceptualisation of the Scientific Revolution, made accessible this time to a public wider than my circle of professional colleagues only.

Now does all this entail a conviction on my part that nothing essentially new was achieved by scientific research after 1700? Or that the new pattern that was established once the mid seventeenth-century crisis of legitimacy had subsided has not undergone any further meaningful change ever since?

Anything but! I join a few economic historians and historians of technology in attaching great importance to distinguishing an eighteenth-century period of 'incubation' between the Scientific and Industrial Revolutions. I am further one of those who find that much of what happened in science in the nineteenth century deserves to be brought under the common denominator of the as yet rather vague concept of 'The Second Scientific Revolution'. However, before

I indicate in a final 'look ahead' what those two episodes of radical change broadly consisted of, I want to recapitulate in a brief 'look back' what steps I have taken in this book to answer the question of how, and due to what, modern science arose in the first place.

THE RISE OF MODERN SCIENCE EXPLAINED: A RECAPITULATION

In effect, I have divided the core question addressed in this book into two parts. How did modern science come about and, once there, how did it survive? The first is a recognised enigma with many proposed solutions, to which I have now added my own. But the second phenomenon is quite as remarkable as the first. The enduring survival of the new nature-knowledge broke with all historical precedent. What we have become used to in our day, the unbroken growth of the scientific enterprise, is the big exception in world history and as such demands an explanation.

To the first question let us reduce the answers given in this book to a list of bald assertions, stripped for the occasion of every residual nuance.

- In two civilisations, the Chinese and the Greek, nature-knowledge was practised in a way that not only provided insights into certain specialised fields but also, in broad strokes, mapped out the whole natural world.
- In China, nature-knowledge took the form of an empirical-practical approach against the background of a broad world-view. In the Greek world, nature-knowledge was more intellectualist and took two different forms: four schools of natural philosophy centred on Athens and, separate from these, an abstract mathematical mode of nature-knowledge centred on Alexandria.
- In principle, the Chinese and the Greek approaches formed equally viable ways of dividing up and coming to grips with the natural world. Only in retrospect can we maintain that it was the Greek approach, and probably not the Chinese, which had the potential (but only the potential, not the certainty) to be turned at some propitious occasion into modern science.
- Whether what was latent would actually emerge depended in the first instance on whether opportunities for transplantation became available.

Transplantation of a certain body of culture from one civilisation to another is one of the great sources of innovation in history. Existing forms and contents can be enriched, and new ones can emerge by way of more or less drastic transformation. But political-military history happened to take a course that ruled out cultural transplantation in the case of Chinese nature-knowledge, while opening up opportunities for it, indeed three times over, with its Greek counterpart.

- Each of those three cultural transplantations, then, was due to military conquest. The first, in Islamic civilisation, occurred in the wake of a civil war which brought the Abbasids to power in about 800 AD. The new capital Baghdad became the centre of a massive wave of translations from Greek to Arabic. The second, in medieval Europe, was made possible by the *Reconquista*, with Toledo as the centre of translation from Arabic into Latin in the twelfth century. The third, in Renaissance Europe, was triggered by the fall of Constantinople in 1453. The original Greek texts were released and translated into Latin first in Italy, and later in the rest of Europe.

- In each of the three civilisations Greek nature-knowledge was received in an atmosphere of enthusiasm and a thirst for knowledge. Translation went hand in hand with enrichment of content. The mental framework as such, however, was retained unchanged. Also the separation between 'Athens' and 'Alexandria' in terms both of personnel and of content remained intact. Medieval Europe deviated in so far as 'Alexandria' fell away and 'Athens' was reduced to just one of its four philosophical schools, that of Aristotle.

- In each of the three transplantation-civilisations, around the edges of the Greek corpus a certain amount of nature-knowledge emerged which incorporated characteristics specific to that civilisation. In Islamic civilisation it was dictated directly by religion, in Europe indirectly. The increasingly extrovert direction that Christianity took in Western Europe was reflected in un-Greek forms of research that were focused on accurate description and practical application. This kind of research remained small-scale and incidental in the Middle Ages but in the course of the Renaissance grew into an independent third mode of nature-knowledge alongside the revived 'Athens' (all four schools) and 'Alexandria'.

- The pursuit of nature-knowledge displays a broadly similar pattern in Greece, in Islamic civilisation and in medieval Europe. In each case an enthusiastic upswing culminated in a Golden Age followed by a sudden, steep decline which did not, however, preclude an occasional late-flowering top performance.

- In Islamic civilisation the individual top performances appeared in the context of partial recoveries from the decline that had marked the end of the Golden Age in around 1050. The recoveries were regional and were confined to Persia under the Mongols, Andalusia under the Berbers and the Ottoman Empire. In each of these three regions there was a fresh revival during which nature-knowledge was and remained oriented towards the Golden Age of several centuries earlier. In contrast, the decline of nature-knowledge in medieval Europe was so thoroughgoing that after its Golden Age and its rapid decline starting in about 1380 there was no revival, not even of an incidental nature.

- Looking back from around 1600 one is struck by two remarkable similarities.

On the one hand, between nature-knowledge in medieval Europe and that of Islamic civilisation after its decline and local revival. In both cases practitioners kept revolving within a closed albeit large circle, and even retrospectively fruitful finds were trapped within that framework.

The other similarity is between nature-knowledge in Renaissance Europe and Islamic civilisation in their initial flourishing periods. In both cases, a similar level of enrichment was achieved across a broad front. In the former case, the Golden Age was brought to an abrupt end by a wave of destructive invasions from about 1050 onwards. Islamic civilisation turned in on itself and fell back on spiritual values, which left little room for nature-knowledge of foreign origin. Europe in contrast was spared invasion. Instead, around 1600 its Golden Age culminated in the first instance in three revolutionary transformations. If we think away the savage invasions of the finely branched Islamic world or even postpone them in our imagination by a generation, a somewhat comparable culmination in the Islamic world begins to look historically conceivable.

- Between 1600 and 1640, abstract-mathematical nature-knowledge begins to link up with the real world by means of validating experiment. In natural philosophy the particles of matter known in ancient atomism are now being provided with explanatory mechanisms within the framework of a partly novel conception of motion. Finally, there is a shift away from unprejudiced, unaided observation, from the search for unknown natural phenomena in the hope of finding a use for them, to systematic, fact-finding experimental research of an even more dedicated, practical bent.

- The first of these three almost simultaneous transformations was the most revolutionary of them all: 'Alexandria-plus' drastically transformed not only the *content* of 'Alexandria' but also its *knowledge structure*. In contrast, the knowledge structure of 'Athens-plus' remained for the time being intact. The third transformation displays, relatively speaking, the least disruptive break with the past. Progressive series of fact-finding experiments were not wholly without precedent, but were now undertaken far more frequently, more intensively and in a more pointed manner.

- Notably, the first of these three near-simultaneous revolutionary transformations, the one pioneered by Kepler and above all by Galileo, might also (albeit of course in a different form) have happened in the Islamic world. It would then have formed the pinnacle of its Golden Age. In contrast, the Islamic world's limited awareness of Greek atomism meant that there was little if any chance of anything like the second revolutionary transformation. And in the absence of an Islamic counterpart to Europe's extroverted urge for accurate, life-like observation, anything like the third transformation may be regarded as excluded from the start.

- What all three revolutionary transformations have in common is that a latent potential becomes manifest. This is the core explanation. But for each of them individually it has also proved possible to point out additional specific causes, each different from the others. And there is the further circumstance that they all took place at the same time. This is closely connected with Europe's not *absolutely* but *comparatively* greater openness, more intense curiosity, greater dynamism, stronger individualism and extroversion, greater readiness to seek salvation in an active earthly existence. All this did not determine, but it did increase, the chance that here, too, a latent potential would this time be turned into actuality.

- The foregoing list of causes is by no means exhaustive. The realisation of a development potential is never preordained or inevitable. The Scientific Revolution need not have happened at all. Or it might have happened later, as a consequence perhaps of a fourth transplantation of the Greek corpus to European settlements in America or Asia.

To the second question of how the three revolutionary modes of nature-knowledge managed to survive, my answers, again in the format of bullet points, are as follows:

- The first period of revolutionary transformation, which lasted until about 1645, provoked an ineluctable, potentially lethal crisis of legitimacy. The philosophy of particles in motion in particular, but also realist-mathematical nature-knowledge, raised many a highly placed eyebrow in view of the apparent strangeness of their tenets and the suspicion of sacrilege that their apparent philosophical-theological implications so readily suggested. Galileo and Descartes became the focus of fierce controversy in Italy, the Northern Netherlands and France. True, civil authorities kept matters in their own hands; no death sentences were imposed and book bans had little effect. Nevertheless, censorship and self-censorship did threaten to take the steam out of the drive to innovate, in a Europe-wide climate of politico-theological hair-splitting and the threat of war of all against all.
- A way out was provided on the continent by the Peace of Westphalia in 1648 and in England by the Restoration in 1660. Both saved Europe from sliding into total chaos and promoted an atmosphere of reconciliation and compromise. This coincided with a rapid shift of the European centre of gravity from the Mediterranean to the Atlantic, economically and politically as well as culturally. The new nature-knowledge was involved in all these changes. This happened notably in the 1660s when under royal patronage two societies were set up, in London and Paris, to promote innovative research.
- In all kinds of ways, the one even more effective than the other, innovative research was neutralised in terms of its possible theological and political implications. One outcome was that experiment (regarded by all concerned as philosophically and politically neutral) was granted a central position in the various investigations undertaken all over Europe.
- That the two leading royal houses went to such trouble (and in France also considerable expense) had much to do with the high expectations held of the new science. It seemed that it could be of great practical use in warfare and in enhancing prosperity. On a much smaller scale it had already happened before 1600, in particular in the service of navigation. But for the time being the chasm between the new approaches and the crafts was far too great to bridge in a single leap.

- This reality remained largely hidden due to the circumstance that, notably in Britain, those high expectations were given expression in an ideology. The Baconian Ideology provided the basis for a belief in the practical use of science on behalf of certain Christian values. It emphasised the most exceptional aspect of Christianity, the possibility offered by Protestantism in particular of seeking salvation by way of dedicated, this-worldly activity. On the grounds of that reason alone, even without any others, the hypothetical achievement of an imaginary Galileo-like figure outside Europe would undoubtedly have run aground.

- The new modes of nature-knowledge did not merely survive but each developed an individual dynamic. The rapid and wide expansion of the realist-mathematical mode came from exploring the complex interaction between mathematical regularity and experimental testing. The natural philosophy of particles in motion was widely adopted due to many attractive features which, taken together, came down to their offering risk-free innovation. Exploratory-experimental research found a dynamic of its own in an ongoing struggle with nature's caprice.

- With the last two in particular, in spite of the presence of or deliberate search for validating criteria, it proved hard indeed to curb the arbitrariness which seemed to lurk round every corner. The best of the second generation sought relief in yet another couple of revolutionary transformations. For the first time since the Greeks, the walls separating the various modes of nature-knowledge started, at least partially, to crumble. The natural philosophy of moving particles, stripped by these men of its dogmatic character, was turned into a potentially fruitful hypothesis. As such, it could be paired up with the realist-mathematical and the exploratory-experimental modes of nature-knowledge, respectively. The first was the work of Huygens and the young Newton, the second of Boyle, Hooke and again the young Newton.

- In both the modes of nature-knowledge that resulted from these transformations, seventeenth-century research reached a number of peaks. But there were also limits to the approach. In both modes it was Newton who ran up against those limits, and who in the end perceived how to overcome them. In the investigations that he recorded in *Principia* and also in *Opticks* he transcended them in (once again) revolutionary ways. In doing so he laid down unprecedentedly strict and partly novel standards which valid scientific research would have to satisfy.

The Scientific Revolution was, first and foremost, a *historical* phenomenon. It was a clearly circumscribed episode, made up of six distinct, both separately and mutually coherent, variously inter-twined revolutionary transformations. With the completion of the sixth one, Newton's, the episode came to a well-marked end. But from the point of view of well-nigh everything else, the Scientific Revolution set in motion a train of events that has yet to cease. With respect to the methods of science, its contents, its professional status, its institutional bearings, its societal preconditions and underpinnings, its capacity to revolutionise the arts and crafts, its expansion world-wide, claims made on its behalf for a monopoly on rational thought or even for leading inexorably to a truly scientific world-view, and also the various countercurrents provoked by science – in all these and many more respects vast extensions and huge changes have kept occurring. I conclude this book with some observations about two further large-scale transformations which, though not quite so radical as the Scientific Revolution, also warrant the use of the epithet 'revolutionary'.

FURTHER REVOLUTIONS: INDUSTRIAL AND 'SECOND SCIENTIFIC'

The vast contrast between *c.* 1600 and *c.* 1700 which I summed up above is not so great in every respect. In particular, the utilitarian effects of the new, mathematical-experimental and fact-finding experimental approaches to the natural world, that is, their loudly advertised capacity to revolutionise the arts and crafts and thus to contribute to prosperity and general human wellbeing, were quite limited. Time and again a large gulf appeared to open up between the theoretical solution of a practical problem (as with Papin's steam-filled cylinder or with Huygens' seaworthy pendulum clock) and one that was truly practicable. In the seventeenth century this gulf was encountered and explored but not bridged. In the next century a range of highly gifted craftsmen of a radically new type emerged, who managed to bridge the gulf at least in their own speciality. Earlier in this book I mentioned in passing the sea-going timepiece with which John Harrison eventually solved the problem of longitude. Of humble descent, and an autodidact in every respect, he won the £20,000 reward set by Parliament due above all to 'his forte, [which] ironically ... proved to be not so much his ingenuity and craftsmanship, great as these were, but his grasp of the principles of horology, which well exceeded that of more learned contemporaries and served as an example to them'.

I also mentioned in passing Newcomen's fire engine based on Papin's idea and first put on the market in 1712. On the one hand, without the earlier insights into atmospheric pressure and the possibility of creating a void space, such an engine could never have been built – a Chinese Newcomen engine at that time cannot even be imagined. On the other hand, the very fact that Newcomen could turn Papin's sketch into a working fire engine and even make it self-acting so that it could maintain its alternating motion far longer than a few strokes at best was entirely due to ingenuity of a kind possessed as a rule not even by the most dexterous of scholars but only by a manual worker expertly familiar with every nook and cranny of his craft. This applies in particular to the various couplings which Newcomen introduced, and to the diverse ways in which, by means of an eduction pipe and the brilliant invention of a snifting valve, he succeeded in getting rid of the condensing water and any remaining air. An even more advanced case of what it meant to be an engineer of this new type appeared half a century later when James Watt transformed the fire engine into a steam engine. Charged with repairing a scale model of a Newcomen engine, he became intrigued by how quickly it quite literally ran out of steam. He began to measure the amount of steam that went into the cylinder with every cycle and so became aware of the enormous fuel loss incurred by the routine cooling off of the cylinder at every stroke of the engine. Watt was lucky that the pioneering investigation of the properties of steam to which this realisation led him took place at the University of Glasgow, where in spite of considerable social distance he had the chance to engage in regular conversation on equal footing with Professor Joseph Black, the premier expert in the world on the subject of heat. To Black he owed awareness of specific heat as a measure distinct from temperature, and somewhat later it turned out that, unbeknownst to each other, both had discovered the phenomenon of latent heat. But further investigations entirely his own led him to the discovery of various other key properties of steam and to a final realisation that, in order to prevent large quantities of fuel from being wasted, the cylinder should be kept close to boiling point, whereas for the creation of a workable vacuum it needed to be cooled down to the temperature of the surrounding atmosphere. Watt would have had every reason to conclude that his investigation had ended in an inherently insoluble paradox which God or Nature had placed in his way to prevent him from improving Newcomen's fire engine not just marginally but at its very core. Instead, he kept mulling it over until, one spring morning walking in the College Green, he hit upon the solution: his famous 'separate condenser', a secondary vessel, kept cool all the time, into which the steam was led to condense after each working stroke without significantly reducing the temperature of the cylinder.

The stories here briefly retold are typical – craftsmen of a new type, trained in natural science or at least its relevant portions, produced the brilliant inventions on which the new tools and machines were based. And virtually all of them were British. Compared with the other leading nation at the time, France, where science was practised with the highest mathematical sophistication then attainable, scientists in Britain used to operate at a level of less refinement and a lower degree of abstraction, which happened to be precisely the level best geared to what was needed in retrospect for practical inventions of the kind just referred to.

Invention, however, could not possibly be the whole story. Without proper funding, even Watt's brilliant find would have remained a scale model or, at best, a machine that here and there pumped water out of mines with enhanced efficiency. Famously, it was the entrepreneur Matthew Boulton who provided the funds required to invest heavily in Watt's machine and to bribe Parliament into prolonging the original patent. The prolongation was indispensable for turning the Boulton & Watt firm which he founded in Birmingham into the hugely successful, monopoly enterprise that it remained until 1800 when the patent expired and the high pressure steam engine took over the market of incipient mass industrialisation. Indeed, quite unlike Watt himself, who had thought of his own device as just a much more fuel-efficient fire engine, Boulton had the vision to assure Watt that it 'would not be worth my while to manufacture for three counties only; but I find it very worth my while to manufacture for all the world', and to act in accordance with that conviction.

In the Introduction I broached the huge, long-standing problem of why it was in Europe, and not any of the other advanced civilisations, where in the decades around 1800 the New World first emerged from the Old. I announced there that I would confine myself to only one, too often neglected part of that problem, the question of how modern science arose, and why it did so in Europe, not elsewhere. Now that we have solved that question and returned briefly to the even larger problem, I want to define where, in my view, the biggest intellectual challenge to its solution resides. On the one hand, there is the emergence in eighteenth-century Britain of the first viable pieces of science-based technology, conceptually prepared during the Scientific Revolution and actually created by engineers of a new type such as (to confine ourselves to the most famous and most significant) Harrison, Newcomen and Watt. There is, on the other hand, a need to invest in these inventions and to market them – invention, after all, needs capital investment and sales efforts to be turned into true innovation. And indeed, economic historians have gone to great lengths to explain how it is that, by the second half of the eighteenth century, the state of the British economy had become such as to yield entrepreneurs both capable and willing to invest on a large scale in new machinery. There was capital enough, there was an unusually skilled labour force and an adequate infrastructure, there was a viable patent law, and the state could be counted on to behave in a sufficiently predictable manner to make arbitrary expropriation at a moment's notice a remote contingency. The big question, then, is to explain how the mutually so different outcomes of two seemingly unrelated, long-term chains of events should coincide both in time (second half of the eighteenth century) and place (Britain). How is it that this world-historically unprecedented business climate, uniquely ready to encourage investment in such unprecedentedly large and powerful machinery as the steam engine, thereby turning technological invention into large-scale economic innovation, came about at the very moment when this new, potentially world-shaking equipment could be and actually was invented for the first time? *Whence, in short, the extraordinary confluence?* Without the Scientific Revolution, no Industrial Revolution, certainly; but also, no Industrial Revolution without a business climate like the one just sketched. It is well known that, due to informal institutions such as the Birmingham 'Lunar Society', the social distance between entrepreneurs like Boulton and scientifically skilled craftsmen like Watt was remarkably narrow for a deeply status-conscious society like eighteenth-century Britain (or really for any other society at the time). But that is about where the question stands at present. Only much closer co-operation between economic historians and historians of science and technology can bring that enigma to a satisfactory resolution.

And now for 'The Second Scientific Revolution'. The bottom line for any viable concept, intended to express in a coherent fashion a whole raft of nineteenth-century changes, seems to be that the pursuit of science begins to catch up with a wide range of modernisation processes set in motion by the Industrial and the French Revolutions.

Take, for starters, what the emergence of science-based technology contributed not only to the first sustained mass production processes in history, but also specifically to the further pursuit of science. The prime mover of the Industrial Revolution itself, the high pressure steam engine, gives rise in the hands of Sadi Carnot to a profound, skilfully abstracted investigation into its fundamental mode of operation, leading to the enunciation of the first basic concepts of thermodynamics. The mass production of textiles (owing, by the way, to a quick succession of inventions which unlike the steam engine involved no science whatsoever) requires, and soon receives, new chemicals for colouring them. This calls into being a chemical industry that soon caters to other markets as well such as, for instance, agriculture and pharmacy. Concurrently, discoveries in electricity and magnetism underlie the establishment of an electrotechnical industry.

In another sense, the Industrial Revolution meant that the Baconian dream was at long last being fulfilled. In our chapter on how the crisis of legitimacy of the mid seventeenth-century pursuit of nature-knowledge was resolved, I invoked, besides various efforts at neutralisation, the rise of the Baconian Ideology. It was that compound of this-worldly Christianity and a belief in progress through revolutionary nature-knowledge that, despite the absence of any tangible product, helped sustain research for the remainder of the century. To what extent this ideology remained, in the course of the eighteenth century, a living resource of legitimacy and sustenance for science in its various aspects is hard to tell. But so much is certain that, largely through the efforts of the French *philosophes*, science acquired new legitimacy by its close association with the liberating movement of the Enlightenment. By the end of the century, with the advent of the Industrial Revolution, a brand-new, truly enduring source of legitimacy takes over – the actual or at least confidently foreseeable attainment, in good part due to science, of unprecedented prosperity for unprecedentedly large, ever-growing sections of the population. In the first chapter of this book I argued that in the absence of the two big modern props of the scientific enterprise – its built-in dynamic and its being widely perceived as lying at the heart of our prosperity – the pursuit of nature-knowledge was bound to display the customary 'boom–bust pattern' which was first broken in the seventeenth-century Scientific Revolution. That was when a built-in dynamic first revealed itself. Early in the nineteenth century it was followed by the other prop – largely science-based technology as the prime generator and guarantee of our modern way of life.

There are more aspects to the notion of 'The Second Scientific Revolution'. It has many ramifications in a Europe-wide phenomenon that has recently been conceptualised under the name of 'history addiction'. There have always been members of an intellectual elite with an interest in the past, including their own, so the expression does not stand for the *rise* of a sense of history as such. Rather, it marks the *democratisation* of the past in the sense of the population at large in around 1800 beginning to realise that it, too, has a history. French Revolutionaries certainly smashed to pieces monuments of a past perceived as wickedly tyrannical, yet they simultaneously perceive themselves to be deeply engaged in the making of history in a radical way, which they see fit to mark by nothing less than a new, rational (because decimally ordered) calendar. Napoleon's conquering armies rob entire nations of large chunks of their cultural heritage and place the stolen sculptures, pictures, archival documents, etc. on display in Paris for the public at large, thus making history truly public for the first time and also fostering in each of the conquered nations a budding awareness of a shared destiny and a shared history. This deepening awareness of a past dimension in almost everything also finds expression in the pursuit of science. Nature itself is perceived as having a history. That is how the history of the Earth becomes a shared concern of numerous investigators. That is also how portions of what was still known in Baconian terms as natural history begin to turn into recognisable biology, leading by mid-century to Darwin's discovery of a phenomenon as deeply rooted in the past as the evolution of species through natural selection. The world now investigated is no longer one created and maintained by a Divine Being, but rather the autonomous product of secular developmental processes.

Another large-scale change in the pursuit of science, due to developments profoundly intensified by both the French Revolution (e.g. the Napoleonic Code) and the Industrial Revolution (in particular, the budding mass production of identical-looking commodities), is *standardisation*. In science, the search for discovery is not so much replaced as, rather, supplemented by disciplined, systematic field-work and by laboratory research carried out by means of standardised methods and calibrated tools. There is also a growing preoccupation with statistics, with units, standards and maps, to the point at which a good deal of research is directed not so much at nature itself as, rather, at the scientific instrument. True, with all the examples just adduced, as also with those that follow, hardly anything is 100 per cent new. In almost every case, the Scientific Revolution and its immediate successor, Enlightenment science, offer incidental precedents. Also, even more importantly, the Scientific Revolution gave rise, as we saw, to many radically novel findings and procedures that have in essence continued up to the present day. The point, however, is that, just as with the First Scientific Revolution, the Second does not display some illusory creation out of nothing but rather a closely intertwined compound of truly revolutionary transformations.

When the French Revolution spreads out over Europe, the predominant, threefold division of natural research into mathematical science, natural philosophy and exploratory-experimental science dissolves. This may look like a rather odd assertion – did I not maintain in the two final chapters of this book that by the 1660s these three modes of nature-knowledge not only began to expand each under its own steam but were also subjected in highly productive ways to unprecedented albeit partial merger? Yes, indeed; but the point is that this remarkably flexible period of partial merger covers some forty years at most and early in the eighteenth century comes to a close. In the larger European powers of the eighteenth century, Britain, France and Prussia, merger ceases and the threefold division once again becomes frozen, this time under the aegis of ever more centralised, royally financed or at least royally sustained institutions. Whereas exchanges between mathematical scientists remain Latinate and international (Euler moves between Berlin and St Petersburg; almost every major European city houses a Bernoulli at some given moment), and whereas experimentation is very much a local affair expressed in the vernacular, natural philosophy gives rise to nationalised '-isms' – dogmatic compounds of ideas, each with some more or less thoroughly adapted science at its centre. Newtonianism is prevalent in Britain, Cartesianism in France and, in Prussia, Wolff's synthesis of key elements of Leibniz's philosophy. Entrenched in national academies, each of the three modes of nature-knowledge (mathematical, experimental, natural-philosophical) is pursued in its own right, until the French Revolution puts things in flux again and new efforts at merger are undertaken everywhere in Europe. One major result is the effective fusion, in the course of the nineteenth century, of what in Thomas Kuhn's parlance were the mathematical and the Baconian (or, in my own terminology, the fact-finding experimental) sciences. Kuhn's signalling of this important event is what moved him in 1961 to speak (albeit rather in passing) of a 'second scientific revolution' for the first time.

Out of the fusion comes a new, disciplinary parcelling up of the sciences, giving rise to astronomy (the only field with a clearly discernible disciplinary history, albeit one that for much of it also included astrology), further to physics, to chemistry, to biology, to geology, etc. Their respective practitioners begin to cluster in communities of their own, which develop a shared identity and set up disciplinary journals. Interested laypeople, engineers, naval and army officers and other non-professionals are summarily banned. Specialist curricula are set up. The sciences break out of the propaedeutic curriculum (as old as the universities themselves) of the arts faculties. Universities begin to perceive a research calling over and beyond their customary teaching role. Besides the universities, museums (open to the public since the French Revolution) and newly established institutions such as testing stations and national laboratories are given a role of their own in advanced research.

The unifying tendency so marked in the near-fusion of the mathematical and the exploratory-experimental sciences yields hosts of scientific syntheses. Some important examples are evolutionary theory, cell theory, bacteriology, thermodynamics with the new concept of 'energy' at its centre, further electromagnetic field theory, the periodic system, astrophysics and physical chemistry. It is precisely this unifying tendency which I started thinking about decades ago when emerging from the seventeenth-century science of music and reading Helmholtz's *Tonempfindungen* (1863; 'Sensations of Tone'), that opened my mind to the idea, vaguely circulating at the time, of a Second Scientific Revolution. The seemingly smooth combination achieved in that seminal book of all kinds of previously disparate elements, coupled with Helmholtz's style of arguing and his way of dealing with scientific instruments, was unlike anything I had encountered in even the best products of seventeenth-century research. This experience gave me an intuitive sense that our present-day science, besides surely being in many important ways the much-expanded product of the Scientific Revolution, is also the product of another, later compound of revolutionary transformations, for which the expression 'The Second Scientific Revolution' certainly seems an apt one. An 'expression' – not yet a concept. What I have listed above is, indeed, a list, with its successive items categorised under a few comprehensive, plausible-looking captions. That there is some mutual coherence between these various items also seems plausible enough; but to define that coherence – that is, to conceptualise what is as yet just a term and a list and a classification schema – is another, much more challenging matter. In the present book (and in its lengthier, profession-directed parent book) the *First* Scientific Revolution has already gone through its first *re*conceptualisation, as well as an explanation that spans several civilisations. But we are still in need of much more analytical work on the *Second*. Not until that further task has also been accomplished will we grasp in a fuller sense how science has become what it is today.

Timeline 1: pre-1600

	China	Greece	Islamic civilisation	Europe
c. 600–400 BC	early text traditions	pre-Socratics		
427–322 BC	early text traditions	Plato, Aristotle		
c. 300–150 BC	development of cosmology	Golden Age		
c. 200 BC–200 CE	Han synthesis			
c. 150 CE	development, refinement	Ptolemy		
c. 800	development, refinement		Abbasids; translations	
c. 900–1050	development, refinement		Golden Age	
c. 1140	development, refinement			Gerard of Cremona in Toledo
c. 1250			al-Tusi	
c. 1200–c. 1300	development, refinement			Albertus Magnus, Thomas Aquinas
c. 1300–c. 1380	development, refinement			Golden Age Middle Ages
1453	development, refinement		conquest of Constantinople	fall of Constantinople; translations
c. 1450–c. 1600	development, refinement			Golden Age Renaissance

Timeline 2: 1600–1700

	Outside events	Pioneers' research: key dates
1592–1610		Galileo in Padua works on a new conception of motion
1600		Gilbert, *De magnete*
1609		Kepler, *Astronomia nova*
1610		Galileo, *Sidereus nuncius*
1613		Galileo, Letter to the Grand Duchess
1616	Catholic Church condemns Copernican doctrine	
1618	start of Thirty Years War	Beeckman meets Descartes in Breda
1619		Kepler, *Harmonike mundi*
1620		Bacon, *Novum organum*
1627		Bacon, *New Atlantis*
1628		Harvey, *De motu cordis*
1632		Galileo, *Dialogo*
1633	trial of Galileo	
1637		Descartes, *Discours de la méthode*
1638		Galileo, *Discorsi*
1639–45	conflict between Descartes and Voet	
1644		Descartes, *Principia philosophiae*
1644–9	Cavendish family exiled in Paris	

1648	Peace of Westphalia	
1652–6		Huygens investigates collision
from 1657	conflict in France about Descartes' ideas	
1660	Stuart Restoration; Royal Society founded	
1661		Boyle, *The sceptical chymist*
1665		Hooke, *Micrographia*
1665–7		Newton's 'wonder years'
1666	Académie founded	
1673		Huygens, *Horologium oscillatorium*
1679		Hooke writes to Newton
1684		Halley visits Newton
1687		Newton, *Principia*
1690		Huygens, *Traité de la lumière*
1704		Newton, *Opticks*

A note on the literature

My grateful thanks are due to Marita Mathijsen and Frans van Lunteren for reading through the entire manuscript and for their critical, unstinting support. I also wish to thank Chris Emery for our constructive and very pleasant collaboration.

Apart from the Epilogue, everything in this book has been dealt with at greater length and with more detailed documentation and theoretical apparatus in my comprehensive study *How Modern Science Came into the World: Four Civilizations, One Seventeenth-Century Breakthrough* (Amsterdam University Press, 2010). That in turn draws in part on my earlier work *The Scientific Revolution: A Historiographical Inquiry* (University of Chicago Press, 1994).

The author whom I have discussed there at greatest length, with a mixture of admiration and criticism, is Joseph Needham. He was the great pioneer of research into pre-1600 Chinese nature-knowledge and of cross-culturally comparative research in the history of science. He wanted to know why it is that modern science did not emerge in China but in Europe. His research resulted in a long series of books under the collective title of *Science and Civilisation in China*. After Needham's death in 1995 the series which had been published from the outset by Cambridge University Press has been continued. Volume IV, part 2, contains a detailed account of Su Sung's water clock. Under Needham's supervision, Colin Ronan published a more accessible series for non-specialists entitled *The Shorter Science and Civilisation in China*. One of Needham's numerous collaborators, Nathan Sivin, together with Geoffrey Lloyd, compared early Chinese nature-knowledge with the Greek in *The Way and the Word: Science and Medicine in Early China and Greece* (New Haven: Yale University Press, 2002). Even so, a chronological survey of Chinese nature-knowledge still has not been written, certainly not in any language accessible to me. It is therefore not yet possible to make any meaningful statements about patterns in its course over time. Incidentally, my brief remarks on how the Han synthesis came about, which render what I found in *The Way and the Word* and in numerous standard works on Chinese history, are almost fully at odds with the views that another sinologist, Kristofer Schipper, who is a Tao master, sketched out in his 2014 translation into Dutch of Confucius' Conversations ('Analects').

There are plenty of surveys of Greek and medieval nature-knowledge. Very useful is David C. Lindberg, *The Beginnings of Western Science* (University of Chicago Press, 1992) (a second revised edition appeared in 2007).

My thinking about the decline of nature-knowledge prior to the rise of modern science owes its initial jolt to Joseph Ben-David's *The Scientist's Role in Society: A Comparative Study* (Englewood Cliffs, NJ: Prentice-Hall, 1971).

Certainly for those who, like me, do not know Arabic or Farsi there is no acceptable scholarly survey of nature-knowledge in Islamic civilisation. That is why I consulted various partial studies and specialised encyclopedias. I refer the reader in particular to *Encyclopaedia of the History of Science, Technology, and Medicine in Non-Western Cultures*, edited by Helaine Selin (Dordrecht: Kluwer, 1997), although not all the entries are equally reliable.

As a rule, nature-knowledge in the Renaissance is not treated as an episode in its own right, distinct from the medieval episode that preceded it and from the Scientific Revolution that unpredictably followed. The closest to it is Allen G. Debus, *Man and Nature in the Renaissance* (Cambridge University Press, 1978).

A great deal of up-to-date information about the Scientific Revolution has been compiled in Wilbur Applebaum (ed.), *Encyclopedia of the Scientific Revolution from Copernicus to Newton* (New York and London: Garland, 2000). Readers who would like to sample interpretations of the Scientific Revolution that are different from mine can do so in a number of recent concise and easily accessible studies such as Steven Shapin, *The Scientific Revolution* (University of Chicago Press, 1996); John Henry, *The Scientific Revolution and the Origins of Modern Science* (Basingstoke: Macmillan, 1997) (third reprint, Basingstoke:

Palgrave Macmillan, 2008); Peter Dear, *Revolutionizing the Sciences: European Knowledge and Its Ambitions, 1500–1700* (Basingstoke: Palgrave, 2001); or Lawrence M. Principe, *The Scientific Revolution: A Very Short Introduction* (Oxford University Press, 2011). Much more detailed is Stephen Gaukroger, *The Emergence of a Scientific Culture: Science and the Shaping of Modernity 1210–1685* (Oxford: Clarendon Press, 2006) (the first instalment of an impressive series). I have briefly summarised and commented on these and several other books on the Scientific Revolution in a 'Postscript' to my historiographical book (www.hfcohen.com under 'Books'). I have also discussed Gaukroger's book in 'Two New Conceptions of the Scientific Revolution Compared' (*Historically Speaking: The Bulletin of the Historical Society*, 14, 2, April 2013, 24–6).

For the individual scholars who together produced the Scientific Revolution I refer the reader to Applebaum's encyclopedia. Much more detailed is the *Dictionary of Scientific Biography*, under the general editorship of Charles C. Gillispie between 1970 and 1980 published in sixteen volumes by Scribner's in New York (meanwhile updated and made available on the internet and as an e-book). For the rest I confine myself here to books dealing with the lives and work of the most prominent revolutionaries. As early as 1948 a splendid biography of Kepler appeared in German by Max Caspar; it has also been translated into English. The best study of Galileo's life and work is John L. Heilbron, *Galileo* (Oxford University Press, 2010). Stillman Drake's translation of the *Dialogue Concerning the Two Chief World Systems – Ptolemaic and Copernican* (Berkeley and Los Angeles: University of California Press, 1953) is unfortunately aimed too much at making Galileo fit the model of a modern scientist. Studies of Descartes vary strongly according to the nationality and professional interests of the authors. French-speaking scholars may emphasise quite different issues from English speakers; similarly so with philosophers and historians of science. I can recommend Desmond M. Clarke, *Descartes: A Biography* (Cambridge University Press, 2006). There is no satisfactory biography of Huygens available in English; but in Dutch Rienk Vermij, *Christiaan Huygens. De mathematisering van de werkelijkheid* (Diemen: Veen Magazines, 2004), is a succinct and reliable account of his life and work. On Newton the standard work remains Richard S. Westfall, *Never at Rest: A Biography of Isaac Newton* (Cambridge University Press, 1980). In 1993 a shorter version without the technical detail of this gripping 800-page book appeared under the title of *The Life of Isaac Newton*. I myself have drawn on Westfall's magisterial work for a study of Newton in Dutch on similar lines to this present book: *Isaac Newton en het ware weten* (Amsterdam: Prometheus / Bert Bakker, 2010). Mention should further be made of Pascal's brilliant letter to Father Noël of 29 October 1647, which is not available in an acceptable English translation but can be found in the original French in every edition of Pascal's works.

Those who want to know 'what happened next', between Newton and the present day, are well served by a quartet of recent surveys of the history of science. Each of them has its own strengths and weaknesses but all are competent, highly readable and with a clear sense of direction. They are James E. McClellan III and Harold Dorn, *Science and Technology in World History: An Introduction* (Baltimore and London: Johns Hopkins University Press, 1999) (the only commendable survey that also covers non-Western developments; a third, revised edition is in press); Frederick Gregory, *Natural Science in Western History* (Boston: Houghton-Mifflin, 2007); Chunglin Kwa, *Styles of Knowing: A New History of Science from Ancient Times to the Present* (University of Pittsburgh Press, 2011); and John Henry, *A Short History of Scientific Thought* (Basingstoke: Palgrave Macmillan, 2012).

Of the more general historical themes in this book the Peace of Westphalia and its effects are well dealt with by Theodore K. Rabb, *The Struggle for Stability in Early Modern Europe* (Oxford University Press, 1975), whereas for recent innovations in the treatment of world history I refer to Patrick O'Brien's summary in 'Historiographical Traditions and Modern Imperatives for the Restoration of Global History' (*Journal of Global History*, 1, 1, March 2006, 3–39). My comments about the comparatively extrovert nature of European civilisation are chiefly based on Max Weber's insights, especially those in his *Gesammelte Aufsätze zur Religionssoziologie* (Tübingen: Mohr, 1920–2). The best analytical work to date on the 'incubation period' between the Scientific and the Industrial Revolution has been done by Donald S. L. Cardwell in *Turning Points in Western Technology: A Study of Science, Technology and History* (New York: Science History Publications, 1972), and by Joel Mokyr in

The Gifts of Athena: Historical Origins of the Knowledge Economy (Princeton University Press, 2002). I have published my own views on the subject from 1998 onwards, most recently so in 'The Rise of Modern Science as a Fundamental Pre-Condition for the Industrial Revolution' (in P. Vries (ed.), Global History. Österreichische Zeitschrift für Geschichtswissenschaften 20, 2, 2009, 107–32). I owe most points about the Second Scientific Revolution to a 'think-piece' by my colleague Frans van Lunteren. Stimulated by the late John Pickstone, who in 2008 organised a roundtable session at a 4-Society meeting in Oxford, Frans and I jointly set up a preliminary workshop on the subject in 2013 (Utrecht), which yielded some valuable insights and was followed by a conference in Leiden (2014). The final book I would like to recommend is on 'history addiction': Historiezucht. De obsessie met het verleden in de 19e eeuw ('History Addiction: The Nineteenth-Century Obsession with the Past'; Nijmegen: Vantilt, 2014) by Marita Mathijsen.

Provenance of quoted passages

p. 5 'thinking with the hands'
Title of *Penser avec les mains*, a book by Denis de Rougemont that
came out in 1936 ('Thinking With the Hands' is also the title of
chapter 7 in my teacher's *magnum opus*: R. Hooykaas, *Fact, Faith
and Fiction in the Development of Science* (Dordrecht: Kluwer,
1999)).

p. 9 'How might what is then perish ... and perishing unheard of.'
Parmenides, verses 19–21 of 'The Way of Truth' (as quoted in
Jonathan Barnes, *Early Greek Philosophy* (Harmondsworth:
Penguin, 1987), p. 134).

p. 22 'guesswork'
Section I:1 in the *Almagest* (G. J. Toomer, *Ptolemy's Almagest*
(London: Duckworth, 1984), p. 36).

pp. 31–32 'For the ancient Chinese ... interdependence.'
In the abridged version of Needham's multi-volume *Science and
Civilisation in China* prepared under his supervision by C. A.
Ronan: *The Shorter Science and Civilisation in China*, vol. I
(Cambridge University Press, 1978), pp. 165, 167–8.

p. 33 'Study was one of several kinds ... to achieve illumination.'
G. E. R. Lloyd and N. Sivin, *The Way and the Word: Science and
Medicine in Early China and Greece* (New Haven: Yale University
Press, 2002), p. 192.

p. 36 'The Chinese cosmos ... readily into play.'
Ibid., pp. 198–9.

p. 46 'a magnificent dead end'

Title of the first chapter of D. S. Landes, *Revolution in Time: Clocks and the Making of the Modern World* (Cambridge, MA: Harvard University Press, 1983).

p. 56 'The work was translated ... the hand of Thabit ibn Qurra al-Harrani.'

As quoted in S. L. Montgomery, *Science in Translation: Movements of Knowledge Through Cultures and Time* (University of Chicago Press, 2000), p. 120.

p. 68 'Thus the Islam of 1300 ... narrow, rigid and "closed" society.'

J. J. Saunders, 'The Problem of Islamic Decadence', *Journal of World History*, 7, 1963, 701–20; 716.

p. 72 'The astronomical science of our days ... not with existence.'

As quoted in R. Arnaldez and A. Z. Iskandar, entry 'Ibn Rushd' in *Dictionary of Scientific Biography* XXII, p. 3.

p. 73 'the master of those who know'

Dante Alighieri, *Divina Commedia*, canto 4, line 131.

p. 76 'intense curiosity about the particulars of nature, most unusual in an age that was forever seeking universals'

Michael McVaugh, entry 'Frederick II of Hohenstaufen' in *Dictionary of Scientific Biography* V, pp. 146–8; p. 147.

p. 80 'The theorems of Euclid ... the delight instilled by our own reading.'

As quoted from N. M. Swerdlow, 'Science and Humanism in the Renaissance: Regiomontanus' Oration on the Dignity and Utility of the Mathematical Sciences', in P. Horwich (ed.), *World Changes: Thomas Kuhn and the Nature of Science* (Cambridge, MA: MIT Press, 1993), pp. 131–68; p. 149.

p. 81 'monster'

Nicolaus Copernicus, *De revolutionibus orbium coelestium* (Nuremberg: Johannes Petreius, 1543), 'praefatio authoris', p. 3.

p. 94 'announced the real existence ... proper knowledge of that world'
S. Shapin, *The Scientific Revolution* (University of Chicago Press, 1996), p. 1.

p. 94 'a diverse array of cultural practices'
Ibid., p. 3.

p. 105 'In trying to prove the Copernican hypothesis ... but of natural philosophy.'
Peter Crüger to Phillip Müller, 1 July 1622; in Johannes Kepler, *Gesammelte Werke* XVIII, p. 92.

p. 116 'The path shall be cleared ... farther-seeing minds than mine shall penetrate.'
Galileo Galilei, *Opere* VIII, p. 190 (*Discorsi*).

p. 122 In both the *Dialogo* and the *Discorsi* he even tried ... progressive series as with his father.
NB: I owe the difference between this statement and its original, Dutch counterpart to my student Sebastiaan Broere.

p. 123 'for courtship as well'
Isaac Beeckman, *Journal* I, p. 228.

p. 125 'What has once started to move moves forever unless it is prevented.'
Ibid., p. 44.

p. 130 'the effecting of all things possible'
Francis Bacon, *Works* III, p. 156 (*The New Atlantis*).

p. 130 'Nature to be commanded must be obeyed.'
Francis Bacon, *Works* IV, p. 47 (*Novum organum* I, aphorism 3).

p. 130 'merchants of light'
Francis Bacon, *Works* III, p. 164 (*The New Atlantis*).

p. 141 'Now that in our times ... narrow limits of ancient discoveries.'
Francis Bacon, *Works* I, p. 191 (*Novum organum* I, aphorism 84).

p. 144 'it being very true ... to have self-esteem first'
Galileo Galilei to Belisario Vinta, 19 March 1610: *Opere* X, p. 298.

p. 144 'if from my youth onward ... to seeking them'
René Descartes, *Oeuvres* VI, p. 72 (*Discours de la méthode*, part 6).

p. 144 'Myself, then, I found to be equipped, more than for other things, for the contemplation of truth.'
Francis Bacon, *Works* III, p. 518.

p. 146 'philosopher and mathematician'
Title page of the *Dialogo*.

p. 155 'wonderful and truly angelic doctrine'
Galileo Galilei, *Opere* VII, p. 489 (*Dialogo*).

p. 160 'The essence and existence ... around his mind.'
As quoted in A. C. Duker, *Gisbertus Voetius*, 4 vols. (Leiden: Brill, 1897–1915), vol. II, pp. xlv–xlvi.

p. 178 'Knowledge is power.'
Francis Bacon, *Works* IV, p. 47 (*Novum organum* I, aphorism 3).

p. 178 'some visible good Work, in the sight of the Multitude'
Thomas Sprat, *The History of the Royal-Society of London* (London, 1667), pp. 365–9.

p. 179 'Seeing the Law of Reason ... the crown of the Law of Nature.'
Ibid.

p. 188 'The cross-sections ... are unequal.'
Benedetto Castelli, *Della misura dell'acque correnti* (Rome, 1628), p. 48 (as reproduced in C. S. Maffioli, *Out of Galileo: The Science of Waters 1628–1718* (Rotterdam: Erasmus Publishing, 1994), p. 49).

p. 191 'the crew's frequent scolding and mockery of this effort to measure longitude in a new way'
Christiaan Huygens, *Oeuvres complètes* IX, pp. 272–91; p. 289.

p. 194 'The knowledge of one single effect ... recourse to experiments.'
Galileo Galilei, *Opere* VIII, p. 296 (*Discorsi*).

p. 195 'To deal with such matters ... experience will teach us.'
Ibid., p. 276.

p. 200 'natural Philosophy, where there is no end of fansying'
Isaac Newton to Robert Boyle, 28 February 1678/9: *Correspondence* II, p. 288.

p. 202 'Ridiculous'
René Descartes to Marin Mersenne, mid January 1630 (quoted from Isaac Beeckman, *Journal* IV, p. 177).

p. 203 'The novelty in the figures ... only 15 or 16 years old.'
Christiaan Huygens, *Oeuvres complètes* X, p. 403.

p. 203 'Mr Descartes had hit upon the way to have his conjectures and fictions accepted as truths.'
Ibid.

p. 204 'has with great ingenuity... pleased with it'
Ibid., p. 406.

p. 212 'The malevolence of inanimate objects ... the charges he intended to collect.'
J. L. Heilbron, *Electricity in the Seventeenth and Eighteenth Centuries: A Study of Early Modern Physics* (Berkeley and Los Angeles: University of California Press, 1979), p. 3.

p. 213 'rather like some enchanted mirror, full of superstitions and ghosts'
Francis Bacon, *Works* I, p. 643 ('De augmentis scientiarum', liber 5, caput 4).

p. 229 'answered pretty nearly'
As quoted in R. S. Westfall, *Never at Rest: A Biography of Isaac Newton* (Cambridge University Press, 1980), p. 143.

p. 231 'Of the Use of Experiments ... duration etc. of experiments.'
As quoted in R.-M. Sargent, *The Diffident Naturalist: Robert Boyle and the Philosophy of Experiment* (University of Chicago Press, 1995), p. 164.

p. 232 'by the intervention ... may at length be made any thing'
As quoted in T. S. Kuhn, 'Robert Boyle and Structural Chemistry in the Seventeenth Century', *Isis*, 43, 1, April 1952, 12–36; 22.

p. 234 'Those [particles] that are of a like bigness ... a kind of Harmony or unison.'
Robert Hooke, *Micrographia* (London, 1665), p. 15.

p. 235 'Female or Mother Principle ... impregnated by the second Principle'
Robert Hooke, *Posthumous Works*, p. 172.

p. 237 'Mechanicall coalitions ... som further cause.'
As quoted in R. S. Westfall, *Never at Rest: A Biography of Isaac Newton* (Cambridge University Press, 1980), p. 307.

p. 237 'Perhaps the whole frame of Nature ... fermental principle.'
Isaac Newton, *Correspondence* I, p. 364.

p. 240 'Struck with joy & amazement'
As quoted in R. S. Westfall, *Never at Rest: A Biography of Isaac Newton* (Cambridge University Press, 1980), p. 403.

p. 252 'ordinary sunlight ... immutable degree of refrangibility'
Isaac Newton, 'New Theory about Light and Colors', *Philosophical Transactions*, 19 February 1672, 3079.

p. 253 'Newton's skill in performance ... an error twice that size.'
R. S. Westfall, *Never at Rest: A Biography of Isaac Newton* (Cambridge University Press, 1980), p. 217.

p. 254 'The Theory wch I propounded ... concluding positively & directly.'

Isaac Newton to Henry Oldenburg, 6 July 1672: *Correspondence* I, p. 209.

p. 254 'the oddest ... in the operations of Nature'

Isaac Newton to Henry Oldenburg, 11 June 1672: *ibid.*, p. 82.

p. 270 'his forte, [which] ... as an example to them'

David S. Landes, *Revolution in Time: Clocks and the Making of the Modern World* (Cambridge, MA: Harvard University Press, 1983), p. 157.

p. 272 'would not be worth my while ... for all the world'

Quotation amalgamated from D. S. L. Cardwell, *Steam Power in the Eighteenth Century: A Case Study in the Application of Science* (London: Sheed & Ward, 1963), p. 62, and from Jenny Uglow, *The Lunar Men: The Friends Who Made the Future 1730–1810* (London: Faber & Faber, 2002), p. 133 (referred to letter of 7 February 1769).

p. 275 'boom–bust pattern'

Stephen Gaukroger, *The Emergence of a Scientific Culture: Science and the Shaping of Modernity 1210–1685* (Oxford: Clarendon Press, 2006), p. 18.

p. 277 'second scientific revolution'

Thomas S. Kuhn, 'The Function of Measurement in Modern Physical Science', *Isis*, 52, 2, June 1961, 161–93; 188.

Index

Printed in the United States
By Bookmasters